SpringerBriefs in Sp

M000187518

Series Editor

Joseph N. Pelton Jr.

For further volumes:
http://www.springer.com/series/10058

Stacey Solomone

China's Strategy in Space

Stacey Solomone
Fairfax Station
USA

ISSN 2191-8171   ISSN 2191-818X (electronic)
ISBN 978-1-4614-6689-5   ISBN 978-1-4614-6690-1   (eBook)
DOI 10.1007/978-1-4614-6690-1
Springer New York Heidelberg Dordrecht London

Library of Congress Control Number: 2013933298

Printed on acid-free paper

Springer is part of Springer Science+Business Media (www.springer.com)

This Springer book is published in collaboration with the International Space University. At its central campus in Strasbourg, France, and at various locations around the world, the ISU provides graduate-level training to the future leaders of the global space community. The University offers a two-month Space Studies Program, a five-week Southern Hemisphere Program, a one-year Executive MBA, and a one-year Master's program related to space science, space engineering, systems engineering, space policy and law, business and management, and space and society.

These programs give international graduate students and young space professionals the opportunity to learn while solving complex problems in an intercultural environment. Since its founding in 1987, the International Space University has graduated more than 3,000 students from 100 countries, creating an international network of professionals and leaders. ISU faculty and lecturers from around the world have published hundreds of books and articles on space exploration, applications, science and development.

Dedicated to:
Phyllis Winick for her unwavering support
Jim Dator for the skills to dream of better
"futures"
Tom Falzarano for the opportunity to write

Foreword

As the United States pivots interest towards Asia, a thorough understanding of China is more important than ever. In this book, Stacey Solomone explores an area of growing power in China, namely, its space industry. However, this book is not a typical rundown of programs and capabilities. The author, instead, focuses on Chinese motivations and intentions within the context of the leadership culture and national identity. What drives the Chinese into space and how far do they want to go? More importantly, what will they do when they get there? Nobody is better qualified to tackle this subject. The author, a Mandarin speaker, capitalizes on first-hand experience in China over the past 20+ years. Who can benefit from this experience? Leaders in the space industry, military professionals, government policymakers, and anyone desiring a better understanding of China should add this book to their library.

Larry D. James
Lieutenant General
United States Air Force

The views expressed are my own and do not reflect the official policy or position of the U. S. Air Force, Department of Defense, or the U. S. Government.

Contents

Chapter 1
Introduction

Questions Abound

Why is China determined to advance so rapidly into space and pursue so vigor-ously every sort of space endeavor imaginable, from sending a probe to Mars to creating its own global navigation satellite system? Why is China so secretive about its space endeavors and technological capabilities? When will China be able to make space an economically viable resource?

Why did China conduct the 2007 direct ascent anti-satellite (ASAT) test, which created thousands of new pieces of trackable orbital debris that, among other things, has endangered the International Space Station? Why would the Chinese aerospace leadership undertake such an act that the global space community has now deemed to be reckless and ill-advised?

Why does China want so desperately to get to the Moon?

How do the people of the Chinese aerospace industry make decisions, and who among them will be the future leaders in space?

These questions, and many more, are addressed in the following chapters.

This book provides new and important information about the many accom-plishments of the Chinese aerospace industry. At the same time, it explores the many problems and challenges that originate from Chinese space initiatives, which, at times, appear overly secretive as these programs are developed and executed. In short, this book seeks to provide a balanced, yet thorough, exami-nation of the technical, political, historical, and sociological aspects of China's goals in space. The intent of the analysis provided in this book is not to critique or criticize the development of China's aerospace endeavors, but to provide a deeper understanding of the Chinese space programs as an organic whole. It is this author's hope that through better understanding, new possibilities for space cooperation between China and Western nations will become possible (Fig. 1.1).

No comprehensive Chinese space doctrine exists that has been published for the rest of the world to see. Aside from three Space White Papers explicitly published to try and alleviate Western concerns over transparency, no doctrine has been put

S. Solomone, *China's Strategy in Space*, SpringerBriefs in Space Development,
DOI: 10.1007/978-1-4614-6690-1_1, © Stacey Solomone 2013

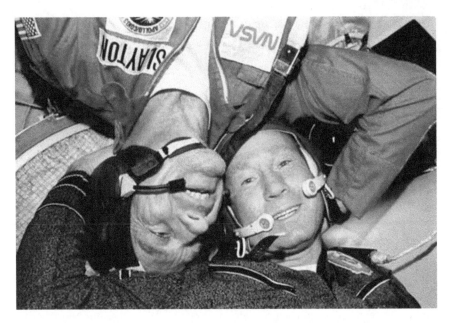

Fig. 1.1 The Apollo-Soyuz mission took place in the midst of the Cold War (Image from http://
www.nasa.gov/pdf/466734main_AP_ST_HG_ApolloSoyuzMission_101509.pdf. Image courtesy
of NASA)

forth similar to the U. S. Presidential National Space Policy or the U. S.
Department of Defense National Security Space Strategy.

The lack of transparency seems both intentional and unintentional. First, it
leaves other space faring nations wondering what China is up to in space—a
purposeful Sun Tze-style tactic of not revealing one's capabilities. Second, many
factors in China's space plans and intentions are implicit, meaning the Chinese
themselves cannot always clearly define their multi-faceted purposes (political,
military, social, historical, technological, and economic), many of which are historically and culturally derived behaviors (Fig. 1.2).

What This Book Is, and Is Not, About

This book explores China in space beyond mere capabilities and hardware technologies. It seeks to examine the *why*, *when*, and *how* of Chinese plans to use space
for strategic advantage and technological advancement. In the following chapters,
you will read about what China's intentions are in space today and in the future. It
explores how space ethics and aerospace age cohorts affect leadership decision-
making on space-related issues and who will be in control of the future decision-
making for the space programs.

Fig. 1.2 Loading the DFH satellite onto a CZ rocket (Image from http://www.cgwic.com/In-OrbitDelivery/CommunicationsSatellite/DFH-4Bus.html. Image courtesy of CGWIC)

This book contains commentary that rarely gets attention when talking about China in space, namely socio-cultural and historical influences on its space plans. In addition, domestic pressures on the space industry are explored. This book seeks to tie China's recent impressive space history to its current activities and plans and then forecast potential future scenarios about where the Chinese aerospace industry may be headed.

In the following pages, you will find tools to help you work with the Chinese in the aerospace industry—rather fresh ways to consider China's historical, political, economic, cultural, and societal factors when engaging with Chinese aerospace professionals. For example, the Beijing University of Aeronautics and Astronautics (Beihang 北航) currently has 26,000 students, including 1,500 doctoral and 6,000 Master's degree students.[1] These and other educational statistics perhaps rightly serve to create an image of a daunting number of future aerospace engineers being produced year after year. Here, we will explore who the future aerospace leaders are, how they think, and how Westerners can best work with them in the future as they rise up the ranks in the aerospace industry.

Understanding China takes years of study and analysis. This book is designed to provide the reader a view of China's aerospace industry from 36,000 feet—how it came about, what it does today, and where it is heading in the future—and should be considered as an appetizer, not the solution to world hunger. This book offers new perspectives and novel approaches to understanding China in space. It also can be used as a template for exploring and understanding China's other mega-industries such as telecommunications, oil, and aviation technologies.

---

[1] For more information on Beihang, visit http://www.buaa.edu.cn.

Although this book attempts to explore China's aerospace industry in a new light, it is not about hardware technologies. There is a chapter on capabilities and some useful appendices, but in such a short book this can only scratch the surface of a very large and significant industry. Nor does this book seek to compare Chinese and American space capabilities, organizations, or missions—although a few observations along these lines are included for illustration and comprehension. Lastly, this book is not all encompassing. So, if an example is given on the historical implications upon, say, the Fengyun meteorological satellite program, this is only one example that should stimulate further thought about similar historical influences that could have played roles in Chinese space goals.

Who Should Read This Book?

The purpose here is to provide the reader with tools to better understand, better communicate, and hopefully better engage and cooperate with Chinese aerospace leaders and professionals on space matters. Therefore, Western political decision makers, military leadership, industrial and military acquisitions professionals, and graduate students interested in China's aerospace industry will benefit from this book. Chinese aerospace leaders, engineers, and scientists who seek to engage more productively with their Western counterparts and understand how, rightly or wrongly, Westerners view them also should read this book. After all, wouldn't you, the Western aerospace professional, like to know what the Chinese think about you?

Lastly, the author wrote *China's Strategy in Space* so that it would be interesting and enjoyable to anybody simply interested in a better understanding of China's aerospace science and technology developments—past, present, and future. For example, readers might want to learn more after hearing a snippet on the news about China sending a probe to the Moon, a Chinese space walk, or the first female taikonaut China recently sent into space. In short, this book seeks to explore China's space programs in terms of its intentions, plans, and limitations. It is the author's hope that readers will finish the book with a sense that not all China does in space is singularly geared to counter the United States in the "space battlefield" and, indeed, that China has strategic plans and visions of its own.

Book Organization

The next chapter, "Why Is China Going into Space?" discusses the historical bases for creating an aerospace industry. Just as the Cold War shaped the U. S. aerospace industry, so, too, did political currents of China's recent history shape the modern capabilities and institutional organization of the industry we see today. Conditions in China were ripe for Chinese leadership to unleash a "tsunami of technological change" that transformed China's nascent space activities into a substantial mega-industry affecting the entire nation.

Chapter 3, "How Is China Going into Space?" first analyzes the hardware (technological capabilities), software (norms and rules to use the technological capabilities), and "orgware" (institutional structure to run the technological capabilities) of the current aerospace industry. The second half of the chapter looks at the orgware's struggle to keep pace with and control new technological capabilities as they are diffused into the space programs and into society. Domestic soft power, especially the power of promulgation, is growing weaker in China. How will aerospace leadership and ultimately the Chinese Communist Party, deal with these impending problems of control? As the Chinese idiom states, Qihu Nanxia (骑虎难下 "Once you ride the tiger, it's hard to get off.").

Chapter 4, "On the Cusp: Innovation's Rise Again," discusses China's grasp for independent innovation in the aerospace industry and how this can be accomplished by embracing indigenous research and development (R&D), exportation of space products and services, limited government financial resources, and the use of dialectic thinking and new managerial methods. The chapter concludes with a look at how successes in independent innovation in the aerospace industry are leading to "sinification" of the global space community.

Chapter 5, "Age Cohorts of the Aerospace Industry," delves into the age cohorts, or generations, of space leaders, engineers, scientists, and technicians. By comparing age cohorts, this tool can provide a better understanding of who the professionals are in the industry, how they think, and why they may take certain actions or make certain decisions in space policy, plans, and programs.

Chapter 6, "Chinese Space Ethics and Decision-Making," discusses the internal philosophies that have played a role in shaping China's current space ethics. The chapter then evaluates how space ethics permeates and influences the decision-making processes, including issues of transparency and the culture of secrecy that continue to be major stumbling blocks during engagement with Western counterparts.

Chapter 7, "The Aerospace Industry and Chinese Society," looks at effects of culture on the domestic aerospace industry and, conversely, effects of space endeavors in Chinese society. The chapter concludes with a small study on Chinese space science fiction and how its unintentional significance illustrates a struggle by China's central government to balance the aerospace industry's need for novelty and creativity with the government's need to control the pace and direction of independent aerospace innovation.

Chapter 8, "Future Scenarios for China in Space," builds upon the previous chapters and offers four distinct potential futures of what the Chinese aerospace industry may look like in a continued growth scenario, a disciplined society scenario, a collapsed society scenario, and a transformational scenario. The purpose of this chapter is to introduce novel ways to think about China's future in space, to foment discussion and communication between the Chinese aerospace professionals and their Western counterparts, and to create a better understanding of

China in space, where it is heading and if Western nations can or will be significant partners in China's journey into space.

Chapter 9, "Top Ten Things to Know about China in Space," points out the importance of the dwindling critical time left to productively engage China in space. Armed with the tools offered below, it is time to engage China in fruitful space cooperation before the opportunity passes by. Or, perhaps even more profoundly, has "The West" already lost the space race, but has not realized it yet? Is China's aerospace industry like the "Borg Mothership" against which all resistance is futile? As Western space organizations hold meetings on Chinese aerospace activities, the same thing is taking place in China as they consider Western space programs, but imagine six times more professionals sitting around the table, most of whom have a respectable command of the English language. 你被强迫要学以前,好好 学汉语吧!

It is not recommended that you read the last chapter without reading the others first. This book was written to offer new tools and suggest new approaches to understand China's goals in space. With these tools outlined in the ensuing chapters, one can better understand, better communicate, and eventually—and hopefully—better cooperate with China in space. Some of the observations that follow are quite candid and perhaps surprising, but, without candor, understanding is not possible. Perhaps the thoughts in the following pages will inspire deeper analysis on this complex topic. 罗倩

Chapter 2
Why is China Going into Space?

An Historical Perspective

The 1966–1976 Great Proletariat Cultural Revolution (文化大革命) has had lasting effects on China's aerospace industry. These impacts can be seen in terms of technological hardware and selected missions, rules and norms in decision-making, institutional infrastructure and organization, and on the aerospace professionals and Chinese people.

The aerospace industry under Mao Zedong (毛泽东) was inexorably tied to the social and political currents of the Cultural Revolution.[1] As a result, residual impacts by the Cultural Revolution remain evident in China's modern space programs.[2] Whereas Mao, and his successor Deng Xiaoping (邓小平), quite successfully utilized the nascent aerospace industry for military and political goals, neither could foresee the momentum China's emerging aerospace industry would gain that helped transform China into a modern society.[3]

Mao took full advantage of the technologies coming from the nascent aerospace industry to support his military and political objectives. However, Mao never fully resolved the contradiction of how to mobilize the masses, support class struggle, and persecute the intellectual class, while also supporting high-technology endeavors for national defense. Therefore, during the Cultural Revolution, a large-scale reorganization of the aerospace programs under the Chinese Academy of Sciences (CAS 中国科学院) was undertaken. Because many of the space-related programs were stripped from CAS and handed over to military control, they were better able to survive much of the chaos of the Cultural Revolution and were partially protected from the Red Guard (红卫兵), who were made up of millions of

---

[1] For more on political considerations for science and technology studies in China, see Ref. [1].
[2] The term "aerospace industry" encompasses China's entire civil, military, commercial, and academic sectors, whereas, the term "space program" indicates one space endeavor such as the human spaceflight program, the lunar program, the Fengyun meteorological satellite program, the Beidou navigation satellite program, etc.
[3] For more on the aerospace industry during the Cultural Revolution, see Ref. [2].

S. Solomone, *China's Strategy in Space*, SpringerBriefs in Space Development,
DOI: 10.1007/978-1-4614-6690-1_2, © Stacey Solomone 2013

red arm-banded militant students across China mobilized by Mao to tear down the heart of Chinese culture, which included institutionalized education and scientific research.[4] The reorganization to turn over many of the space programs to military control alleviated some of the rampant paranoia directed toward CAS and the scientists and engineers that staffed it, since more trust was placed in the People's Army versus the intellectual class. For example, Premier Zhou Enlai (周恩来) placed the China Space Technology Research Institute in Beijing under the People's Liberation Army (PLA 人民解放军) and renamed it the 529 plant to avoid Red Guard attacks on its work.[5] This institute manufactured the Shijian-1 satellite ("Practice-1" 实践一号), which was designed to measure the distribution of the magnetic field in the near-Earth environment.

Also, shortly before China's first satellite launch of the Dong Fang Hong-1 ("East is Red-1" 东方红一号) on 24 April 1970, the entire Dong Fang Hong (DFH) program was handed over to the PLA's 7th Ministry of Machine Building for Astronautics. Other CAS resources were handed over to the PLA's 3rd Ministry of Machine Building for Aeronautics, including the CAS research institutes responsible for developing China's first low-altitude sounding rockets [4]. The reorganization of CAS and placement of control of several institutions under the military provided some protection from the Red Guard, ensuring survival of the aerospace industry. The impacts from those events are reflected in the modern aerospace organization, and, although there have been subsequent reorganizations, the backbone of China's aerospace industry has its roots in the Cultural Revolution.

In addition to the reorganization, Mao and other leaders of the Cultural Revolution attempted to diffuse popularity of aerospace technology achievements by associating them with national pride and deep politico-cultural responsibilities to serve the people. National pride for aerospace endeavors grew from the "Liang Dan, Yi Xing ("Two Bombs, One Satellite" 两弹一星) program, which resonated with the people despite not having a direct technological impact on their daily lives. "Liang Dan, Yi Xing" connected with the Chinese people due to a very concerted public relations effort and, perhaps, in part due to the need for some joy during the chaotic time.

The deep politico-cultural responsibility to serve the people grew from the ancient tradition of the emperor asking the heavens for good weather and good harvests. Therefore, weather monitoring and disaster management, as provided by the aerospace industry, helped fulfill Mao's responsibility to serve the people on multiple levels. To illustrate, in 1968, China began to use foreign satellite weather data. But, to remove any perceived dependencies on foreign technologies, Chinese engineers began developing an indigenous satellite capability for meteorological

---

[4] The students were encouraged to destroy traditional Chinese society to make way for a new socialist China, which resulted in young students turning against their teachers and even their parents, and laid the foundation for deeply seated suspicions and distrust that remain evident in today's society.

[5] In the exhibition room of the 529 plant sits the backup SJ-1 satellite on display [3].

data collection [2]. But the program suffered due to lack of funding and was postponed until the end of the Cultural Revolution, at which time the Fengyun ("Wind Cloud" 风云) satellite program, headed by Weng Jie (翁杰), was resurrected and resulted in the first Fengyun satellite launch in September 1988.[6] Due to the attention given to this mission during the Cultural Revolution, the Fengyun program remains a strong technological accomplishment in the current aerospace industry.

Because mass mobilization and continuous revolution were not sustainable, the Cultural Revolution ended, but the evolution of the aerospace industry did not. There is no doubt China succeeded in building the foundation for a strong and robust aerospace industry during a time when there was very little infrastructure, minimal financial support, and the continuous threat of persecution of scientists and engineers. On the one hand, foreign technologies were no longer entering China, nor were scientists and engineers cultivated to support the growth of space technologies. On the other hand, certain programs of the aerospace industry were financially supported during a time when there were very few economic resources, as long as they were deemed vital to national defense and political objectives. Indeed, the Cultural Revolution had a significant impact on the direction and pace of scientific and technological developments in the aerospace industry and, whether viewed in a supportive or destructive way, one can see effects of the Cultural Revolution in China's modern aerospace industry.

Deng Xiaoping Unleashes a Tsunami of Technological Change

Under the leadership of Deng Xiaoping, the aerospace industry that emerged from the Cultural Revolution underwent another round of changes in direct response to the previous decade. Deng Xiaoping's Four Modernizations, promulgated in 1978 at the Third Plenum of the 11th Chinese Communist Party Central Committee (CCPCC),[7] was a policy designed to end the chaos that reigned during the Cultural Revolution by modernizing agriculture, industry, science and technology, and national defense. Deng's policy created a pragmatic approach toward space sciences and technologies and resolved the place of scientists and engineers in Chinese society, which ended class struggle for the intellectuals.

As a part of the pragmatic approach to science and technology developments, aerospace scientists and engineers were encouraged to leapfrog technologies to

---

[6] For more on the history of the Fengyun satellite program, see Shanghai Academy of Spaceflight Technology's homepage at www.sast.org.cn/index.htm.

[7] For more on decisions made during this plenary session, please see the original proceedings 中国共产党十一届三中全会 on 22 December 1978.

catch up to the Western world. It was also under Deng's leadership that China's aerospace technologies were connected to the people on a practical level. For example, in order to fulfill the deep politico-cultural responsibility to serve the people, space programs were vectored toward communications, education and entertainment, and life science applications. Satellite applications by the aerospace industry into the lives of the people are how the Chinese made a direct connection to space-related technologies that reached beyond the goals of national defense, political objectives, and international prestige.

However, like Mao, Deng, too, could not fully realize that by supporting aerospace endeavors, he unleashed a tsunami of technological change that would shape China and the global space community. Jim Dator coined the phrase "surfing the tsunamis of change" [5] to signify the novelty, magnitude, and power of technologies on societies. Dator draws the metaphor of a tsunami coming from the future that will wipe everything out as it sweeps society into the future.

While the tsunami of technological change comes racing toward the beach, society is having a picnic and is focused on small problems such as the ants in sandwiches, the sand in the drinks, and who forgot to bring the chicken salad. He recommends, instead, taking the attitude of a surfer. Turn around and face the tsunami. Study it and ask other surfers about the conditions of the ocean. Then, wax up the board, plunge in, paddle out, and try to surf it [6]. It is clear that China's space programs are focused on the future and the technological changes aerospace endeavors will bring to society (see Fig. 2.1).

Deng Xiaoping found himself in command of a shattered country when he came to power in the mid-1970s. How was he to heal the economic, political, and social disasters of the Cultural Revolution? What was he to do about a fragmented and dangerous army, and how was he to rebuild a nation? Deng's leadership in the late 1970s to early 1980s, and his vision for aerospace endeavors during this time are partly responsible for unleashing a tsunami of technological change for China's future.

There are six basic reasons Deng fundamentally transformed Chinese science and technology (S&T) with respect to aerospace endeavors. He needed to create an environment in which the aerospace industry could flourish by leading China toward construction of (1) an orderly society, (2) a strong economic base, (3) S&T as the linchpin to the Four Modernizations, (4) a cultivated intellectual class, (5) a modern army, and (6) a solid national educational system.

The Chinese abhor chaos, and yet they suffered through it for a decade. The Cultural Revolution planted the seeds for a shattered, deeply wounded, and paranoid society. Deng's first order of business was to establish order and regain internal stability. As early as 1975, Deng was calling for order, stating, "By putting things in order, we want to solve problems in rural areas, in factories, in science and technology, and in all other spheres.... There was a unit known for its tough and long-standing problems. Its leaders were like tigers whose backsides no one dared touch. Later we made up our minds to spank the tigers... and soon they produced the desired results [7]."

Fig. 2.1 "Our past, our present, our future" (Image from http://chineseposters. net/themes/space-program. php. Image courtesy of IISH/ Stephan R. Landsberger)

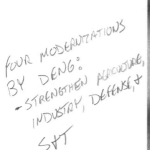

Deng's direction for the country was not based on a future-oriented vision but one of a reactionary nature aimed to reestablish order and retain Chinese Communist Party (CCP 中国共产党) legitimacy. The consequential effects of unleashing the tsunami of technological change likely were not envisioned at that time. He wanted to heal the past and strengthen a weak society. Deng was able to create order out of chaos largely due to societal support for the return of order, which was rooted in an historical precedence. As ancient as the Yi Jing ("Book of Changes" 易经), a text dating back over 4,000 years ago and penned by the mythical Fu Xi (伏羲), the Chinese have viewed adaptation to one's circumstances by following the middle road as a societal value for which to strive, and thus instilling much needed calm back into Chinese society [8]. In ancient China, even the army's objective as a disciplined multitude was to exude a sense of order within society, thus supporting the concept of a People's Army [9].[8]

---

[8] Interestingly, today's PLA struggles with a lack of war fighting experience over the past several decades. Currently, it is gaining some experience through its anti-piracy activities in the Gulf of Aden.

Deng's second order of business was to build a stable economic base. Technologies, including aerospace technologies, played a critical role in economic recovery. During the 1958 Great Leap Forward (大跃进), Mao attempted to rapidly transform the country from an agrarian society into an industrialized socialist economy. Peasants were told to smelt their farming tools and cooking pots in backyard burners set up in the communes to make steel. Trees were cut to fuel the backyard burners, but the fires could not reach ideal temperatures and ended up producing useless lumps of iron. This failed industrialization of China, combined with natural disasters, resulted in the starvation of tens of millions of Chinese people. Right on the heels of such an unimaginable death toll came the calamity of the Cultural Revolution.

Deng's success with economic recovery was no small feat. Deng's Four Modernizations policy launched China into a trajectory it continues to follow today. As Deng stated in 1979, "Our vast territory and rich natural resources are big assets. But many of these resources have not yet been surveyed and exploited, so they do not constitute actual means of production [10]." Deng, thus, was able to get society focused on exploitation of resources to produce a basic economic foundation so that society could weather any future political turmoil or natural disasters.

Thirdly, as Deng stated, "The key to the Four Modernizations is the modernization of science and technology. Without modern science and technology, it is impossible to build modern agriculture, modern industry or modern national defense. Without the rapid development of science and technology, there can be no rapid development of the economy [11]." Of the Four Modernizations, S&T was the linchpin to economic recovery. Had the Cultural Revolution not taken place, there may not have been such a dedicated push toward S&T in the 1980s to recover the country from its dire state. Development of S&T directly led to exploitation and development of China's rich natural resources (land and labor) into productive assets for the country [12]. Thus, this goal fed into the tsunami of technological change.

Deng's fourth goal for China's recovery was directly related to the problem of the banished scientists and engineers, many of whom were "sent down" to the countryside to obtain political re-education. To obtain order and economic recovery, Deng had to support cultivation of aerospace scientists and engineers who would, in turn, help build the S&T base. As a result of Deng's calculated support of the banished intellectual class during this time, today's aerospace scientists and engineers remain strongly organized within the CCP. Deng reintegrated the scientists and engineers back into society by declaring that "everyone who works, whether with his hands or with his brain, is part of the working people in a socialist society [13]." However, by raising their status and material benefits, he also laid the seeds for an emerging middle class of technocrats [14].

Deng encouraged the selection of "several thousand of our most qualified personnel from within the scientific and technological establishment and [creation of] conditions that will allow them to devote their undivided attention to research [15]."[9] Deng's advocacy for wealth distribution, i.e., "to each according to his work," helped to create the current economic class divide. Deng then stated, "There will not be excessive disparities in wealth [16]," perhaps illuminating a miscalculation of growing economic divide, which is perceived in current society as disparities between classes. Deng fomented growth of class status by encouraging class prosperity, which has risen above the standards of living for the majority of the Chinese people. This is a problem President Jiang Zemin (江泽民) exacerbated during his presidency by further elevating the status of technocrats within the CCP. Nonetheless, Deng was very successful in reintegrating scientists and engineers back into society and the CCP; and scientists and engineers in the aerospace industry were able to reclaim their status in society.

The fifth goal was to reign in a wild, undisciplined, and factionalized army. Lin Biao (林彪) and the Gang of Four ("Si Ren Bang" 四人帮) were responsible for the condition of the army in late 1970s. After the Cultural Revolution, Deng was skeptical the PLA was prepared to fight an invading army and was aware of the immediate need to consolidate the military [17]. In addition to leaving aerospace facilities under PLA control, his solution was to encourage scientific research and education within the military and train leading army cadres in modern warfare sciences and technologies [15]. At a meeting of the Central Military Commission in 1975, Deng proposed "peacetime education" for the army. This was an unprecedented action—given that the PLA, founded 1 August 1927, had been established 22 years before the founding of the People's Republic of China (PRC 中华人民共和国).

Prior to the founding of the PRC on 1 October 1949, the PLA had been forged through years of fighting foreign invaders and civil wars. Army cadres "were promoted mainly on the basis of the test of the battlefield [18]." Deng promulgated a radical approach for the PLA Navy, Air Force, and special technical arms of the military to embrace higher-level schooling based on science and technology [19]. Aerospace technologies played an important part of this education, as military officers studied how space technologies could be a force multiplier for ballistic missile technologies; battlefield intelligence (indications and warnings, targeting, and battle damage assessment) via satellite intelligence, surveillance and reconnaissance (ISR) data collection; and beyond line of sight (BLOS) communications via satellites.

Lastly, the sixth reason for Deng's actions that fundamentally transformed the aerospace industry was his work on reforming the educational system of China. During the Cultural Revolution, the country's entire educational system was shut

---

[9] Integrating the intellectual class back into society was accomplished in short order partly due to the Confucian ethic of hierarchical relationships, compiled in *The Analects of Confucius*, and the role of intellectuals in ancient dynastic political systems. This Confucian ethic supported the notion of man's place in an hierarchical society where people's relationships with one another are determined by their status as ruler, subject, father, son, husband, wife, elder sibling, younger sibling, or friend.

down. For 10 years, very little schooling took place across the nation. Deng encouraged reestablishing universities and colleges. He specifically called for reopening the Chinese People's University, reestablishing professional titles for lecturers and professors [20], and eradicating the Gang of Four's labeling of scientists and engineers as the "stinking number nine" (臭老九).

However, Deng was in a rush to see results because society and the economy were in such a fragile state. Therefore, while Deng encouraged the CAS to pay attention to the basic sciences and granted scientists time to work on basic research and teaching in the universities, he also urged them to simultaneously work on applied sciences and technologies [21]. The successes of rapidly reestablishing the educational system were supported by Confucian beliefs in codified education based on a system of rote memorization. Rote memorization stands in stark contrast to establishing independent innovation and is a system that continues to undergo reforms as China's central leadership seeks to find the right balance between rote memorization and creative thinking.

As Deng created a new China based on the six goals mentioned above, all conditions were right to create successful growth for the burgeoning aerospace industry. Deng had, in a very short time, unleashed a torrent of technological change across the country, one in which the aerospace industry could flourish. Deng was successful in restoring Chinese society not only due to support from ancient philosophies but also due to the power of charismatic leadership. In addition to the philosophical foundations of China's ancient ethics-based philosophies of Confucianism and the Yi Jing that supported his efforts, Deng also was adept at using the power of promulgation, a potent form of soft power in China. The Chinese people (老百姓), the army, the intellectuals, and even the Party cadres were in a malleable state during the post-Cultural Revolution. Deng's use of the power of promulgation created motivation for S&T research and development immediately following a time when such endeavors were ostracized (although still peripherally supported for national defense purposes) [22]. Deng was able to accomplish all this by convincing Chinese society that modernity was the answer to recovery and, to accomplish modernity, China had to ride the flood of technological change. Deng used the lure of S&T to offer the Chinese people an alternative to chaos. Research and development into S&T endeavors were transformed into a critical mission—a tsunami to surf—for the entire nation [23].

References

1. Zuoyue, W. 2007. Science and the State in modern China. *Focus* 98(3): 558–570.
2. 张劲夫, "中国科学院与'两弹一星'," 光明网,15日10月2004年; available from http://www.gmw.cn/content/2004-10/15/content_111736.htm.
3. 贾伟伟, "共和国的骄傲-北京卫星制造厂探秘," 人民日报, 31日8月1998年; available from http://rmrbw.net/read.php?tid=1153309.
4. Jim Dator. 1994. "Dogs don't bark at parked cars" (paper presented at the World futures studies federation world conference, Turku, Finland, 23 August 1993).

5. Jim Dator. "Surfing the tsunamis of change" (paper presented at the construction beyond 2000 symposium, Espoo, Finland, 15 June 1992).
6. Shanghai Academy of Spaceflight Technology's homepage at www.sast.org.cn/index.htm.
7. Xiaoping, D. 1984. *"Things must be put in order in all fields"* In *Selected Works of Deng Xiaoping (1975–1982)*, 47. Beijing: Foreign Language Press.
8. The I Ching: Book of changes trans. Richard Wilhelm (New York, Bollingen Foundation Inc., 1950) 179–181.
9. The I Ching: Book of changes trans. Richard Wilhelm (New York, Bollingen Foundation Inc., 1950) 421–422.
10. Xiaoping, D. 1984. *"Uphold the four Cardinal principles"* In *Selected Works of Deng Xiaoping (1975–1982)*, 172. Beijing: Foreign Language Press.
11. Xiaoping, D. 1984. *"Speech at the opening ceremony of the national conference on science"* In *Selected Works of Deng Xiaoping (1975–1982)*, 102. Beijing: Foreign Language Press.
12. Jacques Ellul. 1964. The technological society, New York, Alfred A. Knopf Inc., 1964, 312.
13. Xiaoping, D. 1984. *"Speech at the opening ceremony of the national conference on science"* In *Selected Works of Deng Xiaoping (1975–1982)*, 105. Beijing: Foreign Language Press.
14. Deng Xiaoping. 1984. "Emancipate the mind, seek truth from facts and unite as one in looking to the future." In Deng Xiaoping: *Speeches and writings*, ed. Robert Maxwell, 72–73. Oxford: Pergamon Press Ltd., 1984).
15. Xiaoping, D. 1984. *"Respect knowledge, respect trained personnel"* In *Selected Works of Deng Xiaoping (1975–1982)*, 54. Beijing: Foreign Language Press.
16. Deng Xiaoping. 1984. "Build socialism with Chinese characteristics," 30 June 1984, available from http://www.wellesley.edu/Polisci/wj/China/Deng/Building.htm.
17. Xiaoping, D. 1984. *"The army should attach strategic importance to education and training"* In *Selected Works of Deng Xiaoping (1975–1982)*, 73. Beijing: Foreign Language Press.
18. Xiaoping, D. 1984. *"The army should attach strategic importance to education and training"* In *Selected Works of Deng Xiaoping (1975–1982)*, 74. Beijing: Foreign Language Press.
19. Xiaoping, D. 1984. *"The army should attach strategic importance to education and training"* In *Selected Works of Deng Xiaoping (1975–1982)*, 77. Beijing: Foreign Language Press.
20. Xiaoping, D. 1984. *"Some comments on work in science and education"* In *Selected Works of Deng Xiaoping (1975–1982)*, 71. Beijing: Foreign Language Press.
21. Xiaoping, D. 1984. *"Some comments on work in science and education"* In *Selected Works of Deng Xiaoping (1975–1982)*, 67. Beijing: Foreign Language Press.
22. Lampton, D. 2008. *The three faces of Chinese power: might, money, and minds.* Berkeley: University of California Press.
23. Thomas Kuhn. 1996. *The structure of scientific revolutions* (Chicago: The University of Chicago Press, 1996), xii.

Chapter 3
How is China Going into Space?

Deng encouraged China's scientists and engineers to develop space technologies during a risky time in the PRC's short history. Because China was ripe for change, the aerospace industry was able to build a solid foundation upon which new space programs could be created through the next few decades. A myriad of factors that emerged out the post-Cultural Revolution era laid the perfect foundation for S&T advancements in the aerospace field. With an understanding of why China chose to go into space, it is easier to understand how China created and continues to cultivate a modern, robust, and strong aerospace industry. Science and technology developments under Deng Xiaoping's leadership propelled floundering space endeavors into a modern and admirable technological feat by the start of the twenty-first century.

Hardware, Software, and "Orgware"[1]

The technological achievements by China's aerospace industry consist of three components—hardware, software, and "orgware." The rapid expansion of aerospace endeavors can be examined using these components to shed light on how the complex and dynamic space missions are developing. These three components also can help indicate where the Chinese space programs may (or may not) be heading in the future, and why the *raison d'être* for China's aerospace industry is evolving.

China's rockets, satellites, and ground facilities comprise the technological hardware of China's aerospace industry. The Chang Zheng ("Long March" CZ 长征) family of rockets and the DFH satellite platform are fundamental elements in the modern space programs and will continue to play crucial roles in the future [Fig. 3.1 (Image from http://www.cgwic.com/LaunchServices/LaunchVehicle/LM3B.html) shows a CZ-3B launch]. See Appendices A and B for more on Chinese rockets and functional on-orbit satellite information.

---

[1] For more information on the three components of a technology, see Ref. [1].

Fig. 3.1 CZ-3B launch
(image courtesy of CGWIC)

In addition to the DFH satellite platform, China has developed a means to transport people into space onboard the Shenzhou ("Divine Vessel" 神舟) capsules. China's human space program reached its first major milestone in October 2003 when China launched the Shenzhou V capsule into space with Yang Liwei (杨利伟) onboard, who became China's first "taikonaut"—the Chinese-equivalent term for "astronaut" or "cosmonaut." He successfully orbited Earth 14 times in 21 h before enduring a hard landing 9 miles short of the intended landing site.[2] Since that time, there have been additional taikonaut launches into space that include extravehicular activity, space dockings, and the recent successful trip by China's first female taikonaut onboard Shenzhou IX.

In addition to the rocket and satellite hardware, China has established a complex ground infrastructure to support its on-orbit assets. The three space launch centers (SLCs) within its borders are the Xichang (西昌) SLC located in Sichuan

---

[2] It is interesting to note that the original reports from China stated the landing was only 3 miles from the intended landing site. China waited four years to report the truth. Xi'an Satellite Control Center (XSCC) reported that Yang Liwei lost all means of communicating with XSCC after entering the atmosphere nor could ground-based radars track the capsule. XSCC engineers claimed they had to use cinetheodolites to track the capsule and, based off of that data, they were able to properly control the parachute deployment. For more, see Ref. [2].

Fig. 3.2 CZ-3B launch from
Xichang (image courtesy of
CGWIC)

Province, the Jiuquan (酒泉) SLC in Gansu Province, and the Taiyuan (太原) SLC in Shanxi Province. Construction currently is underway to build a fourth SLC on Hainan Island slated to be completed in 2013 [3]. China's main tracking, telemetry, and control (TT&C) network sites are located at the Xi'an Satellite Monitor and Control Center (XSCC) and Beijing Aerospace Command and Control Center (BACC).[3] In addition, China has several secondary TT&C fixed, ground mobile, and ship-based (Yuanwang 远望) sites. The TT&C network allows aerospace personnel to track satellites, assess satellite bus conditions, make corrections to satellite orbits, and uplink and downlink data and commands to the bus and payload. See Appendix C for locations of China's TT&C sites (Fig. 3.2 Image from http://www.cgwic.com/LaunchServices/LaunchVehicle/LM3B.html).

The rules and norms for use of technological hardware are constituted within the "software" of the aerospace industry. In this context, technological, political, cultural, economic, and historical influences—software—thus define which programs will be pursued and the goals and objectives for each space undertaking. In this sense, the software defines how plans and decisions are made within and among aerospace organizations.

---

[3] The BACC site focuses on human spaceflights.

Influences by the software are generated from the military, civil, commercial, and academic sectors, which make up the overall aerospace industry. For example, the military-industrial complex continues to play a large role in the civilian space programs, perhaps as a residual that was first seen during the Cultural Revolution when many of the aerospace institutes were placed under military control. The norm of the PLA's role in civilian programs was evident in 2003 when Yang Liwei took off for his historic journey. Yang was quoted as saying that he will "gain honor for the People's Liberation Army and for the Chinese nation," [4] thus, clearly stressing the military's role in China's first human space flight.

Understanding these norms and rules (or sometimes referred to as tactics, techniques, and procedures) are critical to understanding how aerospace technologies are and will be used by China. The national social and policy software acts as a bridge between the hardware and orgware. These concepts are discussed further in Chaps. 5, 6, and 7, which outline age cohorts of the aerospace industry, space ethics and decision-making, and the aerospace industry and Chinese society.

The "orgware" of the aerospace industry is comprised of the institutions and authorities formed to control, plan, and coordinate the hardware. In March 1986, aerospace engineer Yang Jiachi (杨嘉墀),[4] along with Wang Daheng (王大珩), Wang Ganchang (王淦昌), and Chen Fangyun (杨嘉墀) [5], proposed the 863 Plan (中华人民共和国科学技术部863计划)[5] approved by Deng Xiaoping to accelerate innovation in China's strategic high-technology R&D fields to support socio-economic and national security needs. To administer the 863 Plan, the Science and Technology Leading Group under the State Council was led by Premier Zhao Ziyang (赵紫阳).[6]

The main ministry under the State Council involved with organizing and unifying long-term space policies and plans is the State Administration on Science, Technology, and Industry for National Defense (SASTIND 中国国家国防科技工业局), formerly known as the Commission on Science, Technology, and Industry for National Defense (COSTIND). SASTIND's primary space-related goals are to coordinate space policy and plans for the State Council.

Under SASTIND, the China National Space Administration (CNSA 国家航天局)[7] was established in 1993 and is responsible to the Premier of China. CNSA was given the responsibilities to help create national space policies; manage development of national space science, technology, and industries; and administer

---

[4] Yang Jiachi, an expert in satellite attitude control systems engineering, made significant contributions to three-axis stabilization on the Fanhui Shi Weixing (返回式卫星) recoverable satellites, and initially worked for CAS before he was moved to CAST in 1968, where he eventually reached the rank of a CAST vice president.

[5] See Ref. [28].

[6] For more on China's Leading Group structure, see Ref. [6].

[7] For more on CNSA, see their website at http://www.cnsa.gov.cn/n615709/cindex.html.

the civilian space programs.[8] In the same year, the China Aerospace Corporation was formed but, in 2000, was restructured into two State-Owned Enterprises (SOEs)—the China Aerospace Machinery and Electronics Corporation (CAMEC 中国航天机电集团公司)[9] and the China Aerospace Science and Technology Corporation (CASC 中国航天科技集团公司).[10]

CASC was given the additional responsibility of handling the aerospace industry's commercial aspects by constructing a new system to accelerate aerospace businesses in space systems, defense systems, applications, and services. CASC employs 140,000 people and has over 130 organizations under it that directly support China's space programs.[11] Some of the major players under CASC are the Chinese Academy of Launch Vehicle Technology (CALT 中国运载 火箭技术研究院),[12] which is China's largest academy working on design, development, and manufacturing of the CZ launch vehicles with over 8,000 engineers and 22,000 employees; the China Academy of Space Technology (CAST 中国空间技术研究院),[13] which makes the DFH satellite platform; the Shanghai Academy of Spaceflight Technology (SAST 上海航天技术研究院), which has over 6,000 engineers and 20,000 employees, and builds components for both the CZ rockets and DFH satellite buses; and the China Great Wall Industry Corporation (CGWIC 中国长城工业总公司),[14] which was established in 1980 as the sole commercial organization authorized by the Chinese government to work with foreign customers to conduct satellite sales, commercial launch services, and international space cooperation (Fig. 3.3 Image from http://www.cgwic.com/In-OrbitDelivery/CommunicationsSatellite/DFH-4Bus.html).[15]

One additional major player is the China Satellite Launch and Tracking Control General (CLTC 中国卫星发射测控系统部), which commands China's TT&C ground infrastructure and is headquartered in Beijing.[16] CLTC falls under the command of SASTIND yet is run by the PLA's General Armaments Department, which is directly subordinate to the PLA's Central Military Commission. The organization and command and control of CLTC illustrate the complexity of the Chinese aerospace industry's orgware and just how interlinked the various aerospace sectors actually are.[17] In short, this example illustrates that the four

---

[8] China National Space Administration, "Organization and Function;" available from http://www.cnsa.gov.cn/n615709/n620681/n771918/index.html.

[9] For more on CAMEC, see http://www.cgwic.com/About/Milestone/Before2004.html.

[10] For more on CASC, see their website at http://www.spacechina.com.

[11] China Aerospace Science and Technology Corporation, "Talented Personnel;" available from http://www.spacechina.com/n16421/n17221/index.html.

[12] For more on CALT, see www.calt.com.

[13] For more on CAST, see their website at http://www.cast.cn.

[14] For more on CGWIC, see their website at http://cn.cgwic.com.

[15] CWGIC, "Company Profile," available from http://www.cgwic.com/about/index.html.

[16] For more on CLTC, see http://www.cgwic.com/Partner.

[17] For more discussion on CLTC's chain of command, see [7] and [8].

Fig. 3.3 DFH-4 satellite bus (image courtesy of CGWIC)

aerospace sectors (civil, military, commercial, and academic) are not easily distinguishable and cannot be separately analyzed.

As is evident from the outline of the above institutions and corporations, China's orgware has a complex structure for its space programs that continues to be reorganized and expanded with new technologies and new missions [9].

To what extent does such a large, and perhaps bloated, bureaucracy impede development of technology hardware and acceptance of new software? Can China's current orgware, as it is now constituted, promote and adapt to technological changes within the aerospace industry? Issues are compounded as China seeks to integrate civil, military, commercial, and academic sectors. The bottom line is that China's goals in space seem to be expanding faster than the capacity of the orgware to manage those goals.

Adaptability of the Orgware

China has a very complex institutional system to unify, organize, and control the planning and development processes for China's space programs. However, this system lacks adaptive vision when it comes to current and emerging space technological hardware. China's aerospace orgware not only has blurred lines of command and control, as seen with CLTC but also is often subject to reorganization due to political objectives more often than for functional streamlining. In fact, China's ministerial-level aerospace institutions have followed a pattern that threatens its continued existence in its current form, namely, that of a large governmental bureaucracy insufficient to adapt to changes in hardware and software innovations.

Ironically, it is the very same orgware that is currently promoting hardware and software innovations within the aerospace industry. Yet, it is stymied in terms of new innovations on the horizon. Therefore, this system ultimately seems to be faced with an adapt-or-die choice.

To illustrate, emerging aerospace technological hardware creates choices where previously there were none [10]. As these new technologies are diffused into civil, military, commercial, academic, and societal applications, the Chinese end users are able to do new things and have new experiences they could not have before with older technologies. This allows the end users to think, plan, and behave differently. Chinese sense of self, others, distance, time, and right and wrong all change as they act in novel ways via these novel technologies.

For example, communications technologies, taken as a whole, began with the written word. It then evolved into the telegraph, telephone, radio, motion pictures, television, computers, Internet, satellites, networks, and smart phones [11]. The medium of communications is evolving at an exponentially fast pace. In China, the rapid development of communications technologies was recognized by Deng Xiaoping as a critical project [12]. Specifically, satellite communications have been used as a source of cohesion for the Chinese people, as it allowed the leadership to reach remote portions of China, provide tele-education and tele-medicine to villages in the inner regions, and promote the application of the Common Language (Putonghua 普通话) in areas where Mandarin previously was scarcely recognizable. However, the CCP is attempting to control the software, or rules and norms, by limiting the content of satellite broadcasting across China to eradicate "excessive entertainment" and promote "traditional virtues and socialist core values" [13].

The same diffusion of satellite communications also has impacted the PLA, which can now more effectively reach BLOS, as is evident in the PLA Navy anti-piracy activities in the Gulf of Aden. In the same vein, new telecommunications opportunities are being created by the commercial sector to exploit satellite-derived communications for profit by offering broadcasting services to an international audience (especially the Chinese diaspora).

In this evolution of technologies, hardware changes first, which allows for new behaviors and activities in the software, as is being witnessed in the communications field outlined above. Yet, orgware lingers behind [14]. Large bureaucracies act in standard fashion to preserve their domains, as established norms are destroyed and reshaped into new modes of behaviors. The orgware typically reacts in one of three ways. It may try to root out the technological hardware and software that is threatening to destroy the old institutional system. But, this is unfeasible once a technology is diffused to the end users. Second, the orgware can disassemble itself and allow for new institutional structure to emerge that is better equipped to deal with new technological hardware and software. This is, of course, highly unlikely, as it is quite rare for an institution to willingly disband itself. Or, third, as is the precarious situation for China's ministerial-level institutional structure for aerospace missions, it can try to make small changes to adapt to the technological changes presented by the hardware and software. Alas, this is only a temporary fix and cannot be sustained. Nevertheless, China's current aerospace orgware has chosen this third option. It has chosen to make minor adjustments in a fleeting attempt to (a) keep pace with the rapid technological changes underway, (b) redefine the nature of those changes, and (c) try to control the changes.

Keeping Pace with Rapid Technological Changes

At the CCPCC 10th Five Year Plan (2001–2005), China's central leadership called for speeding up innovation in China's aerospace industry. The State Council supported the new effort and united space sciences, technologies, and applications under the "China Aerospace Flag." Under this system, the State Council sought to promote higher efficiency to "intensify," "speed up," and "accelerate"[18] aerospace technology development [15]. The new "China Speed" (中国速度) was echoed in the 2006 Space White Paper. It required the aerospace orgware to "make arrangements ahead of schedule, develop frontier technologies, accelerate progress and innovation in space science and technology,... (and) accelerate the industrialization of space activities" [16]. In the field of satellite communications, it specifically called for the aerospace institutions to accelerate commercialization of satellite telecommunications and broadcasting, and develop geostationary orbit communications satellites with longer operating lifetimes, higher reliability, and increased capacity for live broadcast, broadband multimedia, and emergency communications (technological efforts that were well underway).

The 2006 Space White Paper also called for accelerating the buildup of world-class space corporations. During a visit to CASC to discuss the Tiangong-1 ("Heavenly Palace-1" 天宫一号) space lab and the Shenzhou VIII unmanned docking capsule, CCPCC member and Vice Premier of the State Council Zhang Dejiang (张德江) called for accelerating technological innovation and improvements in institutional mechanisms.[19]

This kind of push to develop aerospace technological hardware and orgware at "China speed" will lead to cracks in quality control. During the CCPCC 12th Five Year Plan (2011–2015), the CCP recognized the looming dangers.[20] This Five Year Plan is China's first serious attempt to temper the speed of development by calling for "sustainability" and recognizing that the institutional structure can no longer keep pace with the rapid technological change at those speeds [17].

---

[18] It is interesting to note that in the Chinese language, one often speaks in extremes so that, for example, the weather is *very* hot or *very* cold. This linguistic/cultural phenomenon may be reflected within the aerospace industry's call for accelerated development to a more extreme measure than what would have been found in other cultures in the same situation.

[19] "张德江强调大力推进央企科技创新," 上海证券报, 2011年03月19日, available from http://www.p5w.net/news/gncj/201103/t3504128.htm.

[20] Note, however, the 2011 Space White Paper does not reflect this attempt at cooling off the pace, but continues to promote "China speed."

The CCP has been attempting to focus on a balanced development plan; however, this earnest attempt to slow down the rapid technological change may have come too late. Some might argue that the organization that produces the CZ rocket could not pay sufficient attention to quality control during a time when China was entering a period of intense, fast-paced space launches. On 18 August 2011, due to a malfunction between the devices connecting the servo-mechanism and the second stage Vernier engine, a CZ-2C rocket launched from Jiuquan SLC with a Shijian-11 satellite onboard encountered catastrophic failure [18]. Similarly, others might argue that problems with the Chinese satellites built for Venezuela and Nigeria that experienced malfunctions also reflected too much pressure on outmoded organizations.

CAS has followed suit and has recognized the need for orgware to adjust to hardware and software changes in space technologies. In 2010, CAS wrote *Space Science and Technology: A Roadmap to* 2050 based on China's 2020 S&T plan that was chaired by Premier Wen Jiabao (温家宝). CAS set out to define its role in space science, application, and technology developments through 2050. As an integral institution in China's institutional structure for space development, CAS addressed the lagging orgware problem. "What will be the whole plan for China's space technology?" [19] "We haven't yet done any strategic research over these issues, not even worked out any plans," stated former President of CAS Dr. Lu Yongxiang (路甬祥) [19]. A "new development mode must be shaped.... Achieving this requires us, in the process of China's modernization, to have foresighted overview... [20]." CAS, as China's main consultative think tank on S&T for the nation's decision makers, has recognized that a forward-looking approach involving not only the hardware and software components of aerospace technologies is needed, but a foresighted plan for its organizational structure is also critical.

CAS made the connection between space technologies diffusion into society and adaptation needed by the orgware. "Social changes are without limit... so are the development of science and technology [21]." In the *Roadmap*, issues specifically between the people's interests, science and society, and science and culture are addressed by CAS. However, CAS's quick fix is to tow the central government's "China speed" approach. Specifically for space communications technologies, the *Roadmap* identifies the same short-sighted non-transformational developments it had for the past decades. Namely, the *Roadmap* calls for faster space communication data rates and more key platform technologies that are capable of meeting application needs.[21]

Although CAS recognizes the transformational nature of these space technologies, it does not have the authority to support an orgware adaptation needed to face the emerging aerospace technologies. Arguably, CAS was able to survive the tumultuous times of the Cultural Revolution; therefore, it may be able to survive

---

[21] "Abstract," *Space Science & Technology in China: A Roadmap to 2050,* eds. Guo Huadong and Wu Ji, (Beijing, Beijing Science Press, 2010), 4.

the changes it will face in the near future from emerging aerospace hardware and software technologies. However, CAS has not set itself on a path to sustainability, but instead has chosen a path of clinging on to obsolete institutional structure while the hardware and software rapidly advance—all the while recognizing that CAS is faced with difficult choices. CAS leadership is trying to keep pace with the rapid technological change, but likely will not succeed in its current form based on current decisions for the future aerospace orgware.

Redefining the Nature of Rapid Technological Changes

This coping mechanism is being attempted on two fronts. On the first front, there has been a change in China's approach to developing aerospace technologies. Under Deng Xiaoping, leapfrogging technologies by gaining technologies from foreign countries was touted as the best and fastest way to "catch up." But a shift has begun to make itself apparent, as is evident in the Space White Papers. Whereas leapfrogging of technologies remains a viable and encouraged method to develop aerospace technologies and China continues to closely study foreign space programs, this is not the exclusive path forward.[22] China now recognizes the need to develop technologies indigenously to create a sustainable S&T foundation for the future.

The State Council's plan for an unified system resulted in a reorganization that was not as prolific as the one that took place during the Cultural Revolution, but it did include integration of commercial and academic sectors into the new orgware. In September 2010, Premier Wen stated that China must shift away from the concept of "made in China" to one of "created in China" [22]. China is attempting to accomplish this by breaking down stovepipes and instituting a whole of government approach with respect to space policy and plans. The government is seeking to collaborate with academia and commercial enterprises to remove redundancies in programs by fomenting better information sharing, and to create a more economically viable aerospace industry by encouraging business opportunities for new technologies. There is a long road to travel to overcome stovepiping and compartmentalization within a largely paranoid system.[23] But, efforts are underway. These efforts are reflected in the China Ministry of Science and Technology's (MOST's 科学技术部) goals for S&T programs, which are to "intensify innovation efforts and realize strategic transitions from pacing front-runners"[24] to "...making innovations independently" [23]. The 2011 Space White Paper also supports these efforts. Specifically for satellite communications,

---

[22] 周威,"商业通信卫星市场:综述," 中国航天(第4期2011年): 20–25.
[23] 王军霞, "整体政府的知识协作研究-基于中国的实践," 北京航空航天大学学报(社会科学版) 第23卷第5期(9月2010年) 1–6.
[24] "National High-Tech R&D Program (863 Program)," Ministry of Science and Technology of the People's Republic of China; available from www.most.gov.cn/eng/programmes1/200610/t20061009_36225.htm.

China's aerospace industry will seek to expand value-added businesses, further commercialize and expand the industrial scope, and make new breakthroughs in integration of the space civil, military, commercial, and academic research communities.

On the second front, a large-scale effort is underway to make orgware more suitable for development of advanced space technologies by further integrating societal needs into the industry's goals. When the State Council released China's first Space White Paper in 2000, the goals of China's space activities were in direct support to Deng Xiaoping's policies from the late 1970s to early 1980s. Yet, the State Council recognizes that space technologies exert a most profound influence on modern society [24]. While holistically planning for space sciences, technologies, and applications, the institutional system is playing a very precarious balancing act between sticking to the CCP's approach of "China speed" and attempting to control the pace of diffusion of emerging space technologies into society. The 2011 Space White Paper highlighted the same recognition. "The position and role of space activities are becoming increasingly salient... and their influence on human civilization and social progress is increasing" [25].

How can China's aerospace orgware promote institutional reform to address societal needs, while also "meeting the needs of the state and reflecting its will?" [26] Commercial and academic sectors are being folded into China's space structure in order for aerospace endeavors to promote technologies diffusion into society. The question, however, arises as to how successful will the CCP, State Council, and PLA be in retaining control as they foray into the non-governmental sectors?

Trying to Control Rapid Technological Changes

Chinese central leadership has been exceptionally successful in mobilizing the people through the use of soft power over the past several decades. The successful use of soft power means China is able to get people to behave in ways they desire, not through coercion, but through the artful design of persuasion.[25] To illustrate, Premier Wen has made concerted efforts to connect the aerospace program with the people. After visiting a new base in Tianjin to support CALT's work on the next generation launch vehicle, Premier Wen visited "the masses" and made a point to say that over 200,000 people are living below the minimum subsistence levels—a statement made to garner support for the new high-tech industrial park in Tianjin.[26,27]

---

[25] For more on China's use of soft power, see Ref. [27].

[26] "温家宝在天津考察纪实:把滨海新区建设得更好,"新华网,2010年09月14日;   available from http://news.xinhuanet.com/politics/2010-09/14/c_12552302.htm.

[27] As another example of aerospace industry's work to increase a town's industrial base, see朱国娟, "慈城'三大转变'统筹推进卫星城建设," 区城发展:经济丛刊, (3月2011年) 30–32.

As another example, China's largest national broadcasting network, China Central Television (CCTV 中国中央视台) has sought to export government-sanctioned media directly to the Chinese diaspora and interested foreigners via satellite communications. CCTV News is a 24-hours English language news station broadcast by Asia Satellite Telecommunications Company Ltd (AsiaSat), a Hong Kong-based company, with China International Trust and Investment Corporation (CITIC 中国中信集团公司)[28] and General Electric Company as the two main shareholders. CCTV News broadcasts to more than 85 million viewers in 100 countries and regions, and seeks to give the international audience a window into understanding China.[29] The Chinese private network named Phoenix also has been successful in reaching this large overseas Chinese market.

Despite China's soft power successes in controlling the rapid technological changes, there are some fissures in its control capability. Chinese leaders using aerospace technologies as a means of persuasion have demonstrated only limited successes, including the social and economic changes that accompany the transformational technologies. Whereas once it simply took the power of promulgation from Deng Xiaoping to mobilize the people, currently it is not as authoritative. First, as the middle class within China has emerged, China's central government is finding it harder to mobilize them for their support. Middle class aerospace industry professionals are concerned to see the rich class in lucrative business-oriented industries raise the banner "the sky is high and the Emperor is far away." (Tiangao Huangdi Yuan 天高皇帝远) They and other intellectuals are concerned by what seems to be excessive profit making. The intellectual class, especially within academia, is particularly disillusioned by these perceived inequities.

Second, China's aerospace technologies, such as satellite communications, are enabling media flow throughout all areas of China. For the time, it remains an effective, albeit provisional, tool of the central government. One thing is certain: the current state of the aerospace orgware cannot indefinitely remain in control of these transformational aerospace technologies.

Diminished control is evident in other industries in China. For example, the Chinese Ministry of Railroad's slowness to adapt to modern trains and safety controls was fostered by a number of factors. These included limited decentralization, spiraling expenses, cost cuts, and compromised quality control. The aerospace industry is facing similar issues. The result for the Ministry of Railroad has been train crashes such as the crash on 23 July 2011 in Wenzhou, which subsequently led to protests against the government's lack of safety controls.

Fissures are surfacing in many other industries across China, including the aerospace industry. Indeed, trying to control this rapid technological change may not be sufficient. The ministerial-level institutions responsible for planning China's

---

[28] According to CITIC's homepage, CITIC was originally founded in 1979 by PRC Vice President Rong Yiren with the initiation and approval of Deng Xiaoping. For more on CITIC, visit their homepage at http://www.citic.com/wps/portal/citicen.

[29] "CCTV News, Your Link to Asia;" available from http://english.cntv.cn/20100426/104481.shtml.

space endeavors may need more radical changes to its orgware. Perhaps this aerospace transformation will collapse under its own weight as a new type of technological change emerges. Signs of this may be seen in Kylin TV, which broadcasts 90 Chinese language TV channels to North America, South America, Europe, Australia, New Zealand, Singapore, and Japan via Internet protocol television (IPTV). Kylin TV is dedicated to promoting Chinese language and cultural learning via this new form of television broadcasting potentially threatening to make satellite broadcasting obsolete.[30] Or, perhaps a new institutional system will rise up and take control of the aerospace industry. Imagine a group of Chinese business-oriented commercial satellite companies influencing the current government-centric orgware to work within a new business model. The new Chinese business-oriented aerospace system may even take on the form of a Japanese keiretsu. Only time will tell if China's aerospace orgware has taken sufficient measures to adjust to the future. The next question to ask is, based on the current hardware, software, and orgware of China's aerospace industry, can independent aerospace innovation in China rise again?

References

1. Jim, Dator. 1983. Loose connections: a vision of a transformed society. In *Visions of desirable societies*, ed. Eleonora Masini, 25–45. New York: Pergamon.
2. Manned Mission Threatened by Communication Blackout. 2007. ed. Song Shutao, 新华网, 13 Aug 2007. Available from http://news.xinhuanet.com/english/2007-08/13/content_6523911.htm.
3. China Begins Space Center Construction in Southern Island of Hainan. 2009 新华, 14 Sept 2009. Available from http://www.cast.cn/CastEn/Show.asp?ArticleID=32950.
4. China Claims its Place in Space. 2003 *CNN*, 15 Oct 2003. Available from http://www.cnn.com/2003/TECH/space/10/14/china.launch/index.html?iref=allsearch.
5. Tang, Yuankai. 2009. The Making of China's Own Satellite Navigation System. 北京周报 27, 17 July 2009. Available from http://www.bjreview.com.cn/quotes/txt/2009-07/17/content_207960_2.htm.
6. Carol, Lee, Hamrin. 1992. The party leadership system. In *Bureaucracy, Politics, and Decision-Making in Post-Mao China*, eds. Kenneth Liberthal and David Lampton, 95–124. Berkeley: University of California.
7. China's Space Development. 2008. A tool for enhancing national strength and prestige. *East Asian Strategic Review*, 19–35. Available from http://www.nids.go.jp/english/publication/east-asian/pdf/2008/east-asian_e2008_01.pdf.
8. Harlan, Jencks. COSTIND is Dead, Long Live COSTIND! Restructuring China's Defense Scientific, Technical, and Industrial Sector, *RAND*, 59–77. Available from http://www.rand.org/pubs/conf_proceedings/CF145/CF145.chap5.pdf.
9. Stacey, Solomone. 2006. China's space program: the great leap upward. *Journal of Contemporary China* 15(47): 311–327.
10. Jim, Dator. 1983. Loose connections: a vision of a transformed society. In *Visions of desirable societiesed*, ed. Eleonora Masini, 25–45. New York: Pergamon.

---

[30] "About Us;" available from http://www.kylintv.com/kylintv/page/about-us.

11. Jim, Dator. 2002. Visions, values, technologies, and schools. *On the Horizon* 10, no. 4, 19–24, 2002. Available from http://www.futures.hawaii.edu/dator/education/visions.html.
12. Deng, Xiaoping. 1994. In the first decade, prepare for the second. 14 Oct 1982, in *Selected Works*, *Vol.* 3, 1982–1992. Beijing: Foreign Languages Press.
13. China Limits Entertainment Programs on Satellite TV. 2011. 新华, 25 Oct 2011. Available from http://www.china.org.cn/china/2011-10/25/content_23724679.htm.
14. Jim, Dator. 2006. Alternative futures for K waves. In *Kondratieff waves, warfare and world security*, ed. Tessaleno Devezas, 311–317. Amsterdam: IOS Press.
15. Information Office of the State Council of the People's Republic of China. 2000. China's Space Activities, Nov 2000, Beijing. Available from http://www.china.org.cn/e-white/8/index.htm.
16. Information Office of the State Council of the People's Republic of China. 2006. White Paper on China's Space Activities Published, Oct 2006, Beijing. Available from http://www.china.org.cn/english/2006/Oct/183688.htm.
17. NPC Adopts 12th Five Year Plan. 2011. 新华, 14 Mar 2011. Available from http://www.china.org.cn/china/NPC_CPPCC_2011/2011-03/14/content_22131278_2.htm.
18. Malfunction at Devices Connection Blamed for Orbiter Launch Failure. 2011. 人民网, 6 Sept 2011. Available from http://english.people.com.cn/202936/7589243.html.
19. Li, Yongxiang. 2010. Foreword to *Space Science and Technology in China*: *A Roadmap to 2050*, eds. Guo Huadong and Wu Ji, 9. Beijing: Beijing Science Press, ix.
20. Li, Yongxiang. 2010. Foreword to *Space Science and Technology in China*: *A Roadmap to 2050*, eds. Guo Huadong and Wu Ji, 9. Beijing: Beijing Science Press, x.
21. Li, Yongxiang. 2010. Foreword to *Space Science and Technology in China*: *A Roadmap to 2050*, eds. Guo Huadong and Wu Ji, 9. Beijing: Beijing Science Press, xii.
22. Chinese Premier Stresses Scientific Innovation, Sustainable Growth. 2010. 新华, 13 Sept 2010. Available from http://news.xinhuanet.com/english2010/china/2010-09/13/c_13493083.htm.
23. White Paper on China's Space Activities Published. 2006. State council information office, 12 Oct 2006. Available from http://www.cnsa.gov.cn/n615709/n620681/n771967/79970.html.
24. Information Office of the State Council of the People's Republic of China. 2000. China's space activities, Nov 2000, Beijing. Available from http://www.china.org.cn/e-white/8/index.htm.
25. Information Office of the State Council of the People's Republic of China. 2011. China's space activities in 2011, Dec 2011, Beijing. Available from http://news.xinhuanet.com/english/china/2011-12/29/c_131333479.htm.
26. Information Office of the State Council of the People's Republic of China. 2006. White paper on China's space activities published, 2, Oct 2006, Beijing. Available from http://www.china.org.cn/english/2006/Oct/183688.htm.
27. David, Lampton. 2008. "Minds," in *The Three Faces of Chinese Power*: *Might, Money, and Minds*, 117–163 Berkeley: University of California Press.
28. Ministry of Science and Technology of the People's Republic of China, "National High-Tech R&D Program (863 Program);" available from http://www.most.gov.cn/eng/programmes1/200610/t20061009_36225.htm

Chapter 4
On the Cusp: Innovation's Rebirth

Is China's aerospace industry on the cusp of greatness? Ancient China once made great contributions, such as gunpowder and astronomy, to technology advancements.[1] But since the 1911 fall of the Qing Dynasty, occupation by Westerners, two civil wars, invasion by the Japanese, the Great Leap Forward, and the Cultural Revolution, China had been in a state of turmoil. Several decades of political and military unrest left a once strong environment for basic sciences neglected, and, at times, destroyed.

It was not until the late 1970s under Deng Xiaoping's guidance that Chinese scientists and engineers could safely and effectively begin to rebuild a foundation for basic sciences.[2] Now that the seeds have been sown for innovation to once again take root, will China be able to rise up and take the lead in global aerospace innovations in the future?

China entered into an already well-established international space arena with pre-existing rules, laws, and business norms. China's aerospace industry accepted the circumstances and chose to play by Western rules to gain access into the prestigious aerospace arena. Chinese decision-making in the aerospace industry adapted to, and has since flourished within, the existing international structure. But now, China stands on the cusp of taking a lead role in the global space community and transcending the game of technological catch-up.

Independent Innovation

Crucial to this leap in global aerospace leadership is independent innovation within the orgware. China is showing signs of independent innovation within the domestic telecommunications industry. Might lessons learned from the telecommunications

---

[1] A good resource for China's ancient contributions to the world's science and technology discoveries [1].

[2] "加强科技自主创新 辉煌之后再攀高峰," 人民网, 21 December 2010; available from http://he.people.com.cn/GB/197046/13534748.html

S. Solomone, *China's Strategy in Space*, SpringerBriefs in Space Development, DOI: 10.1007/978-1-4614-6690-1_4, © Stacey Solomone 2013

industry in orgware innovations transfer over to the aerospace industry? China has the opportunity to grasp an entirely new, i.e., Chinese, way to run the domestic aerospace industry and interact with the global space community. This would be something quite foreign to Western ways of doing business, as Chinese aerospace leaders will be able to push their "rules of the road" on the global space markets in both a crowded geostationary orbit and electromagnetic spectrum. China's aerospace leaders are aware of the vast successes in the telecommunications industry by companies such as ZTE (中兴通讯股份有限公司) and Huawei (华为技术有限公司), corporations whose managerial models can be emulated.

Briefly examining ZTE as a model, ZTE is one of the world's leading providers of telecommunications and mobile phone equipment and was founded in 1985 by SOEs associated with China's aerospace industry. With 75,000 employees and 15 R&D centers around the world, ZTE serves more than 500 telecommunications service providers across Africa, Asia, Europe, the Middle East, and the Americas.[3]

ZTE, like the aerospace corporations, entered into an already globally developed industry. ZTE's innovation came from its own development of proprietary technologies, standardization methods, and establishment of a Chinese-style business model in the industry. Historically, in China's telecommunications industry, first generation companies simply imported hardware from foreign companies, such as Motorola. Second generation companies focused on learning sales and marketing techniques. Third generation companies, such as ZTE, placed innovation as a foundational goal for a competitive strategy [2].

ZTE accomplished independent innovation by, first, building in-house R&D centers funded by 10 % of its domestic revenue.[4] Forty percent of its employees work as R&D staff [3]. ZTE then expanded globally by using domestic R&D monies in its international R&D centers [4]. It currently has seven domestic R&D centers (Beijing, Shanghai, Shenzhen, Nanjing, Chengdu, Xi'an, and Chongqing) and eight overseas R&D facilities in the United States (New Jersey, San Diego, and Silicon Valley), France, Sweden, Korea, India, and Pakistan [5]. These overseas R&D facilities allow ZTE to create, develop, and manufacture its own technologies in the telecommunications industry and incorporate explicit and implicit technologies from overseas. Second, ZTE aggressively expanded into the international market with ~70 % of its revenue currently coming from the international market [6]. Third, so as not to hinder innovation, ZTE sparingly uses central government monies for limited R&D support, bank loans, market access, and standardization setting [7]. Fourth, ZTE management is based on the strong Chinese cultural tradition of dialectic thinking, which has supported complex decision-making processes and has introduced innovative managerial methods into the telecommunications industry.

China's aerospace industry is applying the above lessons from the telecommunications industry to accomplish independent innovation. First, with respect to R&D, CASC has eight large R&D and production facilities (CAST, SAST, CALT,

---

[3] For more on ZTE, please see their website at http://www.zteusa.com/about/

[4] "About ZTE;" available from http://www.zteusa.com/about/

Academy of Aerospace Solid Propulsion Technology or AASPT, Academy of Aerospace Liquid Propulsion Technology or AALPT, Shanghai Academy of Aerospace Technology or SAAT, China Aerospace Times Electronics Corporation or CATEC, and China Academy of Aerospace Aerodynamics or CAAA). These R&D centers, plus 12 national key laboratories under CASC, are located in Beijing, Shanghai, Tianjin, Xi'an, Chengdu, Inner Mongolia, Shenzhen, and Hainan. CASC's 140,000 employees are highly educated, with 59 % having obtained postgraduate degrees.[5] CASC has built, trained, and handed over to foreign governments TT&C ground stations to operate communications satellites, but CASC has yet to establish overseas R&D centers as a means by which foreign know-how can be obtained. That said, CASC has branched out to internal academia and commercial sectors to expand its domestic R&D base. The academic and commercial sectors will give CASC much more international exposure. This may, in the near future, lead to adopting a model similar to ZTE by investing monies into overseas R&D facilities or programs. Figure 4.1 below shows the CASC logo—a symbol that is increasingly known and respected around the world.

Second, CASC has made significant progress in exporting products and services on the international market. Beginning in the 1980s, the China-Brazil Earth Resources Satellite (CBERS) program, or Ziyuan in Chinese ("resources" 资源), really paved the way for China to jointly work with international partners in space. China launched CBERS-2B on 19 September 2007, and, most recently, CBERS-3 on 9 January 2012 (which also carried a small Luxemburg payload called Vesselsat-2).

Under CASC leadership, China has exported end-to-end communications spacecraft to Nigeria (the failed NIGCOMSAT-1 launched on 14 May 2007 and the replacement satellite NIGCOMSAT-1R launched on 20 December 2011). It launched the Venezuelan Simon Bolivar satellite, also known as VENESAT-1, on 30 October 2008. This satellite had trouble with solar array deployment and experienced intermittent transmission. China also launched the PAKSAT-1R on 12 August 2011 for Pakistan.[6] All of the satellites were designed, manufactured, and tested by CASC, and launched from the Chinese SLC at Xichang onboard CZ rockets. CLTC provided foreign-based TT&C ground station construction for the customers to operate.

For upcoming satellites, China has signed end-to-end communications satellite agreements with Laos in February 2010 to manufacture, test, launch, and provide the ground segment for the LaoSat-1 satellite[7] and with Bolivia in December 2010 for the Tupak Katari satellite. CGWIC negotiated the contracts, CAST is building

---

[5] "Company Profile;" available from http://www.spacechina.com/n16421/n17138/n17229/c127066/content.html

[6] "Satellite Exports;" available from http://www.spacechina.com/n16421/n17215/n161194/index.html

[7] "LaoSat-1 Program;" available from http://www.cgwic.com/In-OribtDelivery/CommunicationsSatellite/Program/Laos.html and "Tupak Katari Program;" available from http://www.cgwic.com/In-OribtDelivery/CommunicationsSatellite/Program/Bolivia.html

Fig. 4.1 CASC logo

the DFH-4 satellites, CALT will provide the CZ rockets, and CLTC will operate the launches and provide the TT&C ground equipment.

Like ZTE, China's aerospace industry is aggressively pushing into the international market on several fronts, especially illustrated by its ability to launch systems without requiring lengthy clearance procedures. For example, on 31 August 2009, China launched the Indonesian communications satellite Palapa-D, which was built by Thales Alenia Space. Although it experienced a third stage anomaly, delaying its insertion into final orbit, it is functioning on-orbit. On 7 October 2011, China also launched into orbit a Eutelsat communications satellite (Eutelsat-W3C), also based on Thales Alenia Space's satellite bus, marking the first time since 1999 China has provided launch service to a European satellite operator.[8] Most recently, on 6 December 2011, CGWIC and Thales Alenia Space announced it had reached an agreement to launch by 2014 the National System of Satellite Communications (NSSC) for Turkmenistan.[9] See Appendix D of this book for a list of all foreign payloads launched by China.

Third, whereas China's major telecommunications corporations are able to limit government involvement, CASC remains an SOE. That said, this does not mean CASC leadership does not recognize the need for controlled market freedom. By the end of 2009, CASC had total assets at Renminbi (RMB) 153.3 billion (b) and a total profit of over RMB 7b.[10] CASC's mission statement reflects this sentiment as

---

[8] "Chinese Rocket Sends French Telecom Satellite into Space," 新华, 7 October 2011; available from http://www.spacechina.com/n16421/n17212/c152361/content.html

[9] "CGWIC Signs Turkmenistan Satellite Launch Services Contract," 06 December 2011; available from http://www.cgwic.com/news/2011/1231_揭牌.html

[10] "Company Profile;" available from http://www.spacechina.com/n16421/n17138/n17229/c127066/content.html

it strives to support the spirit of Chinese space business and entrepreneurial culture.[11]

CASC's ventures to generate domestic revenues are encouraged by the central government to alleviate some of the central budgetary constraints. CASC's social responsibility activities in domestic economies allow them to exercise limited market freedom from the central government monies. CASC is focusing on making space profitable to local economies, with large-scale construction projects underway to build industrial parks in cities such as Tianjin. The local economies provide opportunities to sell space-related services and spin-off technologies, but more importantly, Chinese space industries provide a means to grow local industries, create jobs, and increase the domestic standards of living.

As a part of poverty relief efforts by the aerospace industry, from 1994 to 2000, CASC took part in a CCPCC poverty alleviation program designed to lift 80 million people out of absolute poverty.[12] CASC leadership recognizes that its social responsibility will pay off in terms of profits down the road. As more industrial parks are built, providing jobs for local economies, CASC is able to expand its consumer base for products such as personal navigation devices and satellite communications services. For example, CASC and China Telecom (中国电信集团公司), China's third largest mobile telecommunications provider and an SOE, signed a five year agreement in August 2009 in which CASC would implement favorable space segment resources for navigation and communications so that China Telecom can construct a robust 3G network. In return, China Telecom will provide CASC with comprehensive information services.[13]

On 11 January 2012, China Telecom Europe signed an agreement with Britain's Everything Everywhere mobile operator to launch mobile services in the UK. This service is specifically designed to target Chinese residents and businesses in the United Kingdom.[14]

As another illustration, in 2011, Beijing NavInfo Science and Technology Company, Ltd. acquired Mapscape from the Netherlands for RMB 61 million (m). The Beijing-based NavInfo has 60 % of the domestic market share for in-vehicle navigation maps and 50 % of the domestic market share for GPS-based phone navigable maps. Through this acquisition, China will gain advanced

---

[11] "中国航天事业之魂——航天'三大精神'," 29 November 2006; available from http://www.spacechina.com/n25/n144/n206/n216/c76674/content.html

[12] "CASC Commended for Poverty Alleviation Achievement," 8 December 2011; available from http://www.spacechina.com/n16421/n17219/c185366/content.html

[13] "China Telecom Signs Five-year Strategic Agreement with CASC," 25 August 2009, available from http://www.chinawirelessnews.com/2009/08/25/9336-china-telecom-signs-five-year-strategic-agreement-with-casc/

[14] "China Telecom Will Launch Mobile Services in UK," 11 January 2012; available from http://www.chinawirelessnews.com/2012/01/11/10853-china-telecom-will-launch-mobile-services-in-uk/

navigation technologies.[15] This exemplifies how the Chinese SOEs recognizing the need for limited domestic free market can use innovative negotiation methods to potentially squeeze out free-of-charge GPS usage, such as personal mobile navigation or computer-generated mapping and imaging services as found on Google Maps, out of China and dominate domestic satellite navigation and communications services. Revenues from these types of domestic services will alleviate some of the dependencies on central government monies.

Fourth, like ZTE, CASC is embracing the cultural tradition of dialectic thinking which supports complex decision-making processes and innovative managerial methods with Chinese characteristics. CASC currently is constructing a new managerial system designed to accelerate development of space and defense systems, applications, and services.[16] For example, CASC Party Secretary for Scientific Development Plans Li Jixiang (李继祥) is pushing for innovation in managerial style and corporate culture to foment R&D breakthroughs while also keeping the CCP's role relevant in innovative managerial systems. He was able to create a complex system whereby seemingly disparate ideologies (sensitivity to the CCP's involvement balanced with implementation of innovative thinking practices that are outside the norm) work together to complement each other.[17] This indicates CASC's acknowledgement for more management innovation while retaining a role, at least de jure, for the CCP. For example, information stove-pipes are a recognized problem in the aerospace industry; and duplication, unequal distribution of resources, and lack of knowledge sharing among engineers inflates costs [8]. Aerospace leaders realize the precarious balancing act needed to encourage innovation, and, with it, potential for failures, while also relying on the CCP for oversight to help eliminate some of the stovepiping between space programs. Yet, integration of civil, military, commercial, and academic sectors also threatens national security issues related to space endeavors as more transparency occurs.

CASC, like ZTE, is on the cusp of truly independent innovation. CASC leadership realizes they cannot continuously leapfrog technologies because this means they will have to follow foreign strategies in the global space community. China does not want to continue to follow and play S&T catch-up, but is striving to learn to innovate and choose its own path to space. Just as China's Dawning Corporation's Chairman Li Guojie (李国杰) of the Shuguang-1 (曙光一号) supercomputer stated, "Our spirit of innovation is to avoid following the same route as the global industry leaders. We will never catch-up with large multinationals if we follow their strategies. You can't leapfrog when you are following others in the same direction" [9]. China not only recognizes the limits of leapfrogging stages of technological development but also sees using overseas R&D centers, self-

---

[15] "NavInfo Acquired Holland's Mapscape for $9.3 million," *Aerospace China* (Spring 2011): 21.

[16] "Company Profile;" available from http://www.spacechina.com/n16421/n17138/n17229/c127066/content.html

[17] "李继祥:科学谋划发展打头阵," 2 July 2011; available from http://www.spacechina.com/n25/n142/n154/n188/c94196/content.html

generated revenues, and indigenous management techniques as a basis to grasp emerging technologies.

Sinification of the Global Space Community

Because the Chinese have a long and deeply held belief that central to the world is their civilization,[18] deep-seeded values exist that are rather Sinocentric in nature and can be found within China's modern values and ethics. China's rise in the global space community will facilitate bringing these Chinese values and ethics into the international community and will challenge Western norms. Below are three aspects of how China has begun to specifically affect the global space community through a Sinocentric view of the world (also known as sinification).

First, China is quite adept at acquiring foreign aerospace technologies and creating new and improved models by recombining them in novel ways and sinifying them rather than developing new technologies from scratch. Coined "recombinative innovation," [10] China is able to acquire foreign space technologies and incorporate them very successfully into its satellites. China, aware that with recombinative innovation comes both explicit and implicit knowledge, has been extremely successful in sinifying the explicit knowledge so that implicit influences are reduced. Even President Hu Jintao (胡锦涛) recognized the successes of recombinative innovation when he stated, after the successful launch of Chang'e-2 lunar orbiter, that the aerospace industry had become quite adept at boosting recombinative innovation through importation, absorption, assimilation, and re-innovation integration.[19]

To illustrate, launched in October 2005, the Beijing-1 Earth observation satellite was built by the United Kingdom's Surrey Satellite Technology, Ltd (SSTL) in cooperation with Beijing Landview Mapping Information Technology Company, Ltd.[20] Based on SSTL's advanced microsatellite bus and at a cost of RMB 101m, Beijing-1 offers 4 m resolution imagery to the Chinese central government for disaster management.[21,22]

---

[18] The emperor represented the center of the world that was made up of concentric rings of civilization. The further away from the center, the more barbaric one was.

[19] 胡锦涛, "在庆祝探月工程嫦娥二号任务圆满成功大会上的讲话," 新华网, 20 December 2010, available from http://news.xinhuanet.com/politics/2010-12/20/c_12900124.htm

[20] "SSTL 150," available from http://www.sstl.co.uk/Downloads/Datasheets/SSTL_150-Feb-09

[21] "Beijing-1 Satellite Starts Remote Sensing Service," 新华, 8 June 2006; available from http://english.peopledaily.com.cn/200606/08/eng20060608_272262.html

[22] To note, SSTL and Chinese subsidiary of BLMIT, Twenty-First Century Aerospace Technology Company, Ltd. (21AT), signed a contract in June 2011 for SSTL to build three new electro-optical satellites based off the original Beijing-1 design. The new DMC-3 constellation will provide images at 1 m resolution. The deal was part of a larger trade agreement signed by Premier Wen Jiabao and Premier David Cameron worth ∼RMB 13.9b. This

Beijing-1 touts modularization and has reconfigurable onboard hardware and software.[23] Modularization is a concept the Chinese adopted and are pursuing with the goal to provide future on-orbit services in which broken modules within the satellite bus would be replaced in situ by a separate on-orbit robot launched specifically to service the aging or defunct satellite.[24] As another example of Chinese recombinative innovation, the Shenzhou manned capsule, similar to the Russian Soyuz, was facilitated by a 1995 Russian technology transfer agreement. However, the Shenzhou has been made distinctly Chinese in design and is wider, longer, and has more mass than the Soyuz.[25]

Second, Chinese-style negotiations are not new to Western businesses; however, understanding the differing negotiation styles between China and foreign space partners and/or adversaries is still in its infancy. For example, ProtoStar, Ltd., which launched the ProtoStar-1 communications satellite on 7 July 2008, to a slot at 98.5°E geostationary orbit, was accused of emitting interfering signals with satellites owned by the Chinese military and AsiaSat, Ltd. While ProtoStar, Ltd., claimed it had International Telecommunication Union's (ITU's) confirmation that it was operating within ITU rules and regulations,[26] and claimed that China did not show evidence that ProtoStar-1 was actually causing interference,[27] AsiaSat, Ltd. insisted ProtoStar-1 emissions were unauthorized.[28] For months, China was successful in bogging down the litigation process. ProtoStar, Ltd. finally declared bankruptcy[29] and auctioned off its two satellites (ProtoStar-1 and -2) to Intelsat,

---

(Footnote 22 continued)

contract offers a new business model in which the DMCii parent company SSTL will own the spacecraft and lease the capacity to China, similar to business models used for telecommunications but not for remote sensing services. Jonathan Amos, "China and UK Strike Space Deal," BBC News, 29 June 2011; available from http://www.bbc.co.uk/news/science-environment-13946179

[23] "British Company Delivers Beijing-1 EO Satellite," available from http://www.sst-us.com/news-and-events/2009-news-archive?story=522

[24] For more on recombinative innovation efforts on Chinese satellite buses, see Ref. [11]

[25] Dean Cheng, "Chinese Space Capabilities," 17 October 2003; available from http://csis.org/files/attachments/031016_chengppt.pdf

[26] "ITU Confirms ProtoStar's First Satellite is Operating Fully within All Regulatory Guidelines," ProtoStar Press Release, 24 October 2008; available from http://www.protostarsat.com/news/24oct08.html

[27] "November 2008 Satellite Finance Article and Interview with Philip Father, ProtoStar's President and CEO," ProtoStar News Story; available from http://protostarsat.com/news/nov08.html

[28] "Media Statement," AsiaSat, 23 September 2008; available from http://www.asiasat.com/asiasat/EN/upload/doc/pressrelease/news_20080923.pdf

[29] "Declaration of Cynthia M. Pelini in support of ProtoStar's Chap. 11 Petitions and Various First Day Applications and Motions," Case No. 09-12659, 72668-001\DOCSDE:1511092; available from http://delawarebankruptcy.foxrothschild.com/uploads/file/ProtoStar%20Declaration.pdf

Ltd. on 29 October 2009.[30] The Chinese aerospace industry has a unique negotiation style that Western partners must not only understand, but one in which they
also must be willing to work within, to establish productive and cooperative space
relations.

Third, China is pursing aggressive standardization methods within the domestic
aerospace industry and global space community. Chinese aerospace leaders are
looking for new methods to assess S&T work within a system of existing laws,
regulations, guidelines, and norms. Domestically, MOST is attempting to socialize
a new system to institutionalize standardization for new technologies independent
of bureaucratic biases and political favoritism [12]. CASC is in the initial phase of
trying to standardize knowledge management while recognizing there still is a
long way to go on this front [13].

Although China faces an uphill battle to implement its norms and standards
within the global space community due to its tardy entrance into the arena, it is
making headway. For example, China is seeking to include its domestic Beidou
(北斗) satellite navigation system into global civil aviation applications and
standardization regulations [14]. China approached the International Civil Aviation Organization (ICAO) to incorporate the future Beidou global navigation
satellite system (GNSS) into ICAO. China and ICAO agreed that standards and
interoperability issues would be the best place to start discussions to eventually
develop a new system that would incorporate Beidou into international air
navigation.

CASC is focusing on both domestic and international standards,[31] and in 2011,
established the China Academy of Aerospace Standardization and Product
Assurance to develop aerospace standards and quality assurance.[32] According to
Deputy General Manager of CASC Yuan Jie (袁洁), CASC already has embraced
the push for Chinese standardization to raise its international influence and have a
competitive and authoritative voice in the international arena.

Like CASC, CAS is on the same path to standardize its efforts and make them
the international norm. The proposed Digital Earth Scientific Platform and the
Earth System Simulation Network Platform are a part of China's Earth observation
megaproject, in accordance with requirements by the CCPCC's 11th Five Year
Plan (2006–2010). Working in conjunction with CASC, CAS expects the com-
bination of the two platforms to result in technological breakthroughs for antici-
pated energy resource shortages and ecological deterioration. The megaproject,
China's high-resolution Earth Observing System (EOS),[33] is scheduled for testing

---

[30] "Intelsat Named as Successful Bidder for ProtoStar 1 Satellite," Press Release, 30 October
2009; available from http://www.intelsat.com/press/news-releases/2009/20091030.asp

[31] 袁洁,"实施航天标准化工程支撑航天科技工业新体系建设," 中国航天报 (19 August
2011); available from http://www.spacechina.com/n25/n144/n206/n214/c67094/content.html

[32] "China Academy of Aerospace Standardization and Product Assurance Established,"
Aerospace China (Spring 2011): 23.

[33] China's Earth Observing System (EOS) should not be confused with NASA's Earth
Observation Satellites, which has the same acronym.

by 2013 and is expected to be fully operational by 2020. This megaproject seeks to combine space-borne data from China's Fengyun, Huanjing, and CBERS satellite constellations with airborne data; and import data from the National Aeronautics and Space Administration (NASA), the National Oceanic and Atmospheric Administration (NOAA), and the European Space Agency (ESA). CAS envisions controlling the data via network nodes in Beijing, Kashi, and Sanya; and adding four overseas data reception stations and network nodes by 2050 to become an international multi-discipline network platform, scientific data exchange source, and broadcast center [15]. When complete, China's EOS is expected to provide continuous, all-weather observation coverage of the entire Earth.[34]

Standardization will, in the near future, provide China with a bargaining chip [16] in the global space community—a factor the Western space-faring world will struggle with as it slowly comes to realize that Chinese values and norms must not only be recognized but also embraced in future international aerospace dealings and negotiations.

Plans for increased sinification have already begun in China's space programs. Despite China not being asked to participate in the International Space Station, it independently undertook construction and launch of a space lab, with near-term plans for a space station. The Tiangong-1 space lab, which is the first phase to developing a full-time manned space station, is a landmark of China's capacity to build an innovative and independent human space program.[35] As China continues to work hard to accelerate innovation in aerospace technologies, law, ideology, professionals, culture, and politics [17]; it is plausible that the future space station will be used by China as another bargaining chip in international negotiations for manned access to space.

However, much threatens to derail China's ambitious space plans. Without leadership from a charismatic personality such as Mao or Deng, the central government is losing its ability to wield domestic soft power in the form of leadership promulgations to effectively control the direction of the prolific space projects. As the aerospace engineers and scientists make headway in space technologies, it may become more difficult for the power of promulgation to redirect their work that is, perhaps, in line with political goals but not with scientific or technical objectives. The Chinese central government's tight control over the direction of the space programs illustrates the perpetual struggle of orgware for control and its lag behind innovation. Also, China's educational ethos of Confucian rote memorization is working against its potential to innovate and create new thinking, new approaches, and new technologies in the aerospace industry. Plus, China's mounting social, financial, environmental, and legal concerns loom large over the CCP. It may become difficult to continue to financially prop up the bureaucratic space programs

---

[34] "China's High-Resolution EOS Expected to be Fully Operational by 2020," 19 March 2012; available from http://www.spacechina.com/n16421/n17212/c214032/content.html
[35] "温家宝会见天宫:号指挥部成员;贺国强等:同会见,"新华网 (30 September 2011); available from http://politics.people.com.cn/GB/1024/15799846.html

with an uneven payoff to society. Lastly, it is not without historical precedence that China's central leadership has made a move against technology as a way to retain power, such as during the Boxer Rebellion. The next three chapters examine aerospace "software" in terms of the people who make up the Chinese aerospace industry, space ethics, decision-making processes, and societal influences—all of which have significant impacts on the rules and norms (software) of this tsunami of technological change.

References

1. Joseph, Needham. 1965. Science and civilisation in China. Cambridge: Cambridge University Press.
2. Fan, Peilei. 2010. Developing innovation-oriented strategies: Lessons from Chinese mobile phone firms. *International Journal of Technology Management* 51(2–4): 183.
3. Fan, Peilei. 2010. Developing innovation-oriented strategies: Lessons from Chinese mobile phone firms. *International Journal of Technology Management* 51(2–4): 184.
4. Liu, Xielin. 2010. China's catch-up and innovation model in IT industry. *International Journal of Technology Management* 51(2–4): 214.
5. Fan, Peilei. 2010. Developing innovation-oriented strategies: Lessons from Chinese mobile phone firms. *International Journal of Technology Management* 51(2–4): 186.
6. Liu, Xielin. 2010. China's catch-up and innovation model in IT industry. *International Journal of Technology Management* 51(2–4): 213.
7. Liu, Xielin. 2010. China's catch-up and innovation model in IT industry. *International Journal of Technology Management* 51(2–4): 214.
8. 李睿,"论航天型号项目的协作及项目群的管理," 航天器环境工程 27:4 (8 月2010年):519–521.
9. Zeng, Ming, and Peter Williamson. 2007. *Dragons at your door: How Chinese cost innovation is disrupting global competition*, 65. Boston: Harvard Business School Publishing Corporation.
10. Zeng, Ming, and Peter Williamson. 2007. *Dragons at your door: How Chinese cost innovation is disrupting global competition*, 79. Boston: Harvard Business School Publishing Corporation.
11. 郭振伟, 黄威, 朱北园, 冯永, 遇今 "通过'再设计,再分析,再验证'提升关键单机产品成熟度," 航天工业管理 (6月2011年): 54–59.
12. 王再进, "国外科技评估的历史,现状及其启示," 北京航空航天大学学报(社会科学版) 19:4 (12 月2006 年): 76–80.
13. 魏新让, "标准化视角下知识型企业的知识管理初探, "航天标准化2 (2011): 7–11.
14. 秦智,"北斗卫星导航系统:民航应用标准国际化的机遇和挑战,"中国航天6 (2011): 10–14.
15. Development Strategy and Roadmap for Space Science, Applications and Technology to 2050. 2010. In *Space Science and Technology in China: A Roadmap to 2050,* eds. Guo Huadong and Wu Ji, 42–94. Beijing: Beijing Science Press.
16. Zhan, Ailan, and Zixiang Tan. 2010. Standardisation and innovation in China: TD-SCMA standard as a case. *International Journal of Technology Management* 51(2–4): 465.
17. 易杰雄, "创新, 加速现代化的最佳选择,"北京航空航天大学学报(社会科学版) 15:1 (3 月2002 年): 1–5.

Chapter 5
Age Cohorts of the Chinese Aerospace Industry

Within China's aerospace industry exist several age cohorts,[1] or generations, of professionals. These aerospace scientists, engineers, and technicians of the same age cohorts share commonalities in experiences, ethics, and ways of thinking. People who grew up in the same time span can have similar beliefs and thought processes due to shared historical events (wars, revolution, or economic depressions), child rearing practices, technologies, societal trends, and career paths that shape their personalities in similar fashion.

China's aerospace professionals have shared personal experiences through China's recent and volatile history from celebrating the founding of the PRC to embracing the technological revolution. As one cohort ages and a new cohort comes to power, new beliefs, values, and ethics may result in shifts in decision-making processes, actions, and policies.

On average, age cohorts change every 10–25 years. Although age cohort analysis does not and cannot define every person within a cohort, it can be used as one of several tools to better identify and understand the aerospace professionals' potential trends in decision-making actions and processes. Table 5.1 summarizes the eight generations of professionals in the Chinese space program that have, for the purpose of this analysis, been labeled "age cohorts."

Generations of the Current Aerospace Professionals

China's 1st generation of aerospace professionals can be characterized as capable and driven leaders. Early in their careers, they received Western exposure while studying abroad, helping give them an appreciation for the world outside a newborn PRC. They also built their careers on very successful military service (such as Marshall Nie Rongzhen (聂荣臻), born in 1899, who served in the 8th Route Army, and Qian Xuesen (钱学森), born in 1911, who served as a Colonel in the

---

[1] Strauss and Howe [1].

S. Solomone, *China's Strategy in Space*, SpringerBriefs in Space Development, DOI: 10.1007/978-1-4614-6690-1_5, © Stacey Solomone 2013

Table 5.1 Age cohorts of aerospace professionals

Age cohorts	Years born
First generation	b. 1890s to 1911
Second generation	b. 1900 to 1920
Third generation	b. mid-1920s to mid-1930s
Fourth generation	b. 1940s
Fifth generation	b. early to mid-1950s
Sixth generation	b. early to mid-1960s
Seventh generation	b. 1970 to 1976
Eighth generation	b. 1980s to 1990
Ninth generation	b. mid-1990s to 2000

These age cohorts are not meant to coincide with CCP leadership generations

These categories are rough estimates and, of course, there will always be professionals who may associate with the former or latter age cohort based on their personal life experiences or preferences

US Army). As military leaders, these professionals rose to become heads of their fields in the 1950s. Marshall Nie rose to lead China's nuclear weapons program and Qian rose to lead China's rocket and satellite programs.

Aerospace professionals of this age cohort were willing and driven to push for what they believed was best for China at that time. They were skilled survivors and thrived despite any major setbacks. First generation aerospace professionals can be characterized as intelligent, strategic thinkers, loyal to a newly born China, and politically savvy, which enabled them to survive the Cultural Revolution chaos. These shared characteristics were formed by the culture of the time which, in turn, shaped these aerospace warriors into the leaders they became. Their personalities reflect shared beliefs and ethics of the aerospace community of this age cohort.

The 2nd generation of aerospace professionals, such as Yang Jiachi, who was born in 1919 and was one of the founders of the 863 Plan, is marked by protection from their patron Premier Zhao Ziyang, also born in 1919. Professionals of this 2nd generation were born around 1900–1920 and grew up in times of civil wars and foreign invasion. This exposure during their formative years gave them their drive to succeed in the aerospace field, perhaps as a coping mechanism or as an escape from the depravity that they grew up in.

Despite the civil wars and foreign invasions, as young engineers in the 1930s and 1940s, they were afforded opportunities to study abroad for higher education in both the Soviet Union and the West. In mid-life, they witnessed the birth of a new China and then radical communist land reforms in the 1950s, which led to the Great Leap Forward and subsequent starvation of tens of millions of their compatriots. Tempered by tragedy, they are reformists at heart, but were unable to fully express their vision due to a weakened patron-client relationship under Zhou Enlai and Zhao Ziyang, who held secondary status under Mao's hardened rule. Therefore, as middle-aged aerospace professionals, they found themselves delving into their work and making great technological strides for the Motherland despite limited financial and institutional support from the central government.

The "nose-to-the-grindstone" approach to life is the way this age cohort was able to survive. Their endeavors were justified in their minds as a way to boost China into the world's space community. For example, Yang Jiachi's contributions to the FSW satellite program were very valuable to Mao's desire to use the FSW imagery for political gains. Despite their work often being interrupted due to ebbs and flows in political struggles, these careerists contributed to creating a solid foundation for the aerospace industry. Later in life, these professionals were awarded for their contributions to China and likely found themselves in consultancy positions within CASC or CAS as senior advisors for China's current and future space programs. Similar to Zhao Ziyang's deeply felt connection with the young college students protesting in Tiananmen Square in 1989, these elder 2nd generation aerospace professionals also took on the role of mentors to the young aerospace engineers of the 6th generation. To illustrate, as Yang Jiachi stated, "Scientists always act as psychological coaches in the selection of astronauts, encouraging them to be fully confident."[2]

The 3rd generation of aerospace professionals, born between the mid-1920s and mid-1930s, are made up of scientists and engineers who were afforded higher education in technical fields either domestically or in the Soviet Union. Having grown up during a slightly more stable time in early China, they tend to be quiet and tow the party line—a survival mechanism developed in their mid-lives as they suffered through the Cultural Revolution. Because of this factor, they evolved into a recessive age cohort just getting by and working on their projects. Although this age cohort was not constituted of visionaries or strategic thinkers, they did however make significant contributions to China's space programs. For example, Ouyang Ziyuan (欧阳自远), born in 1935, joined the CCP in 1956 likely with pride and hope for the new country. While his career was just beginning to take off in geochemistry, he was able to re-direct his efforts to research effects of underground nuclear testing, which likely saved his career and possibly his life, during the tumultuous years of the 1960s and 1970s. Later in life, he was again able to re-direct his career in geochemistry to focus on the Chinese lunar program.

Inspired by collisions of meteorites on the Moon's surface,[3] he lobbied the central government to mine the Moon for helium-3, used in nuclear fusion technologies, and was later selected as Chief Scientist for the China Lunar Exploration Program (CLEP中国探月). Other aerospace professionals of this age cohort include Wang Yongshi (王永志), born in 1932, and Liu Jiyuan (刘济源), born in 1933. Both received graduate-level education in the Soviet Union during the early 1960s giving them exposure to foreign thinking and approaches. Wang rose to become Chief Designer of the Shenzhou V and VI modules and, like other members of his age cohort, is adept at conforming to the political line in order to

---

[2] "Two Astronauts to Spend a Week in Space," 新华, 4 June 2004; available from http://www.china.org.cn/english/2004/Jun/97313.htm.

[3] "Ouyang Ziyuan's Moon Dream Coming True," *China Daily*, 26 July 2006; available from http://china.org.cn/english/scitech/175923.htm.

maintain his career. Liu, who joined the CCP in 1952, rose in the ranks of the aerospace's military-industrial complex. Like Ouyang, he was able to survive the Cultural Revolution due to the military nature of his work during that time. Liu was the Deputy Director of the 7th Ministry of Machine Building Industry (Astronautics), which contributed directly to Mao's "Three Bombs, One Satellite" effort. Professionals of this age cohort received some protection from Zhou Enlai and Nie Rongzhen partly by being placed under PLA control and partly by being relocated to the Third Front. Professionals of the 3rd generation are particularly adept at moving up the CCP ranks as they seek to please Party patrons to secure their careers.

Members of the 4th age cohort, born in the 1940s, are the current top leaders of the aerospace industry. Leaders such as President Hu Jintao, born in 1942, and Premier Wen, born in 1942, grew up during civil war and foreign invasion and entered adulthood during the Cultural Revolution. People of this age cohort tend to favor "red" over "expert" as they witness the rise of the younger generations and their perceived lack of moral understanding. CAS President Liu Yongxiang (路甬 祥), born in 1942, has had a very successful career and illustrates the precarious balancing act current technical aerospace leaders must perform between being "red" and "expert" to better their careers.[4]

The 4th generation of aerospace professionals understand CCP leadership's professed need for political rectitude, and they stand ready to toe the political line. President Hu was particularly aggressive in cultivating "morality" and increasing political virtues of the younger scientists, engineers, and technicians. The 4th generation currently in charge of the space programs sees the moral compass of their successors as going awry and feels it is their duty and obligation to redirect younger aerospace professionals back onto a correct path of development, not due to a lack of technical expertise but rather lapses in "communist values." By promoting those who are "red," the current aerospace leadership is grooming future leaders who are less technical but can play the political game. This may become a problem later in China's aerospace endeavors as truly skilled strategic thinking space engineers may be overlooked for promotion in lieu of those willing to grasp on to older values to get promoted.[5] While 4th generation aerospace professionals are able to grasp the big picture, their focus remains on building a good image of the CCP at the expense of sidelining experts to lower rank positions, thus risking a weakened aerospace industrial base in the future.

The 5th generation age cohort, born roughly in the 1950s, can be characterized as being good stewards of the status quo. Though their careers may be truncated as 6th and 8th generation future aerospace leaders may bypass them in the 2020s, they are driven by ideology and remain highly nationalistic in support of a "sinicized" view of the world, sometimes referred to as Han nationalism.

---

[4] To view a full bio on Lu's educational and professional careers, see http://www.chinavitae. com/biography/Lu_Yongxiang/full.

[5] Lam [2].

Aerospace leaders of the 5th generation, such as Chen Qiufa (陈求发), born in 1954, and Sun Laiyan (孙来燕), born in 1957, do not tend to hold as many shared values as the previous generations. While they do tend to yearn for the old ways of life and value systems of their predecessors, there is not a cohesiveness found among them, as can be seen in preceding and succeeding generations.

For example, Sun Laiyan, CNSA director from 2004 to 2010, is a highly educated technocrat and received an engineering Ph.D. in Paris.[6] On the other hand, Chen Qiufa, of the Miao Minority and current CNSA director who replaced Sun in 2010, is a CCP loyalist who moved up the SASTIND ranks while maintaining strong Party affiliations. Wan Gang (万钢), born in 1952, current Minister of Science and Technology and a Vice Chairman of the Chinese People's Political Consultative Conference (CPPCC), holding both positions since 2007, is the only current cabinet member who is not a CCP member. With diverse backgrounds and lack of a shared future vision or values-based cohesiveness, this age cohort does not have strong bonds between them. So, whereas the previous age cohort was very driven for the good of the motherland, this group tends to fall back on old familial ties to their local hometowns. Chen, a Hunan native, sought to deepen defense industrial R&D cooperation with Zhou Qiang (周强), governor of Hunan,[7] which included advances in aerospace technologies. Wan, a Shanghai native, returned to his hometown and held the position of president of Tongji University from 2004 to 2007.[8] Sandwiched between the headstrong 4th generation and the rising 6th generation, this aerospace age cohort is often overshadowed by these other age cohorts.

Finally, there are the emerging aerospace professionals who will be taking over the direction of China's aerospace industry in the near future. These are the 6th, 7th, and 8th generations—all of whom vary greatly from their predecessors. It is these age cohorts who may, in turn, greatly alter the projected trajectory of the space programs.

The space professionals of the 6th generation, those born in the early to mid 1960s, are now in their 40s and poised to take leadership roles around 2020. They were born into hardship and likely experienced deprivations and tragedies during the Cultural Revolution. But, enduring the chaos of the Cultural Revolution as children galvanized them as an age cohort. As the chaos ended and China began to rise again, they savored the sweet taste of victory, albeit tempered by the losses during childhood. This generation is made up of technocrats with can-do attitudes and a common positive vision for China's future in space.

---

[6] "领导介绍: 孙来燕;" available from http://www.cnsa.gov.cn/n615708/n620168/n677011/46560.html.

[7] 徐蓉, "周强会见来湘考察陈求发一行," 湖南日报, 26 May 2005; available from http://news.163.com/10/0526/10/67JQ5BP400014AEE.html.

[8] To view a full bio on Wan's educational and professional careers, see http://www.chinavitae.com/biography/Wan_Gang/full.

However, they cannot fully realize their vision due to unwillingness to take political risks due to fear. Their experiences as children have resulted in a generation seen today as lacking fresh ideas and innovation, yet they are politically reliable because they seek to avoid political conflict. In the 1980s, they were pragmatic and proud as they emerged as young adults from the Cultural Revolution. This resulted in the successes of recombinative innovation in their aerospace endeavors. They found they could play it safe yet still make strides in the space programs by learning from foreign space programs such as SSTL and the Soyuz mentioned earlier. Interestingly, this generation is driven by nationalistic goals despite having been born into the Cultural Revolution. Most of China's taikonauts were born into this age cohort—Yang Liwei (杨立委), born in 1965, Zhai Zhigang (翟志刚), born in 1966, Nie Haisheng (聂海胜), born in 1964, Fei Junlong (费俊龙), born in 1965, Jing Haipeng (景海鹏), born in 1966, and Liu Boming (刘伯明), born in 1966.[9]

While once making great strides via recombinative innovation, as elders, the 6th generation of space professionals may become followers due to limited independent innovation experience. For example, Zhang Qingwei (张庆伟), born in 1961, who became general manager of CASC in 2001 and director of SASTIND in 2008 and 2009, received a domestic education and went up the ranks within the aerospace industry, mostly at CALT, in the mid-1980s and 1990s.[10] Because he lacked exposure to the global space community, he attended the 1992 summer course at the International Space University. But, those of the 6th generation aerospace industry face a glass ceiling when it comes to promotion due to President Hu promoting "red" over "expert." Helping to break through the ceiling, currently, Zhang Qingwei is gaining political experience in his home province as Acting Governor of Hebei and Deputy Secretary of the CCP Hebei Provincial Committee, thus becoming competitive with China's Communist Youth League (中国共产主义青年团) politicians of the same age cohort. Zhang joined the CCP in 1992, became an All-China Youth Federation member in 2007, under which the Communist Youth League falls, and is currently a member of the 17th CCPCC (see footnote 10).

Like his fellow aerospace professionals, members of this age cohort must make extra effort to demonstrate their loyalty to the Party and political trustworthiness by becoming members of the CCP or joining the Communist Youth League. If not, they face hitting the glass ceiling in their careers as technical experts while their compatriots who are experts in ideology and propaganda are able to continue to rise up the aerospace ranks to leadership roles.

---

[9] Of exception are two of the three taikonauts onboard Shenzhou-IX. Whereas Jing Haipeng is China's first repeat taikaonaut, his team also included Liu Wang, born in 1969, and China's first female taikonaut Liu Yang, born in 1979.

[10] To view a full bio on Zhang's educational and professional careers, see http://www.chinavitae.com/biography/Zhang_Qingwei/full.

As the Cultural Revolution came to an exacerbated end, the Lost Generation was born. China's 7th generation of aerospace scientists, engineers, and technicians came into a country without an institutionalized education system. In their formative years, they were taught the thoughts, poems, and songs of Mao Zedong while living in a commune-based society, perhaps even being separated from their parents as little children.

The members of China's Lost Generation are society's reminder of the shame of China's recent history, and the aerospace professionals are not immune to these societal biases. Not only do they represent the generation born into the chaos of the 1970s when Cultural Revolution zealotry reached its peak, they also are the first to be born into the one-child policy, which was a hard adjustment for a family-oriented Confucian society. Society has turned away from them, and the central government is actively overlooking grooming them for future leaders and, instead, concentrating on cultivating younger aerospace professionals. This has resulted in the Lost Generation being rather ambivalent to their lots in life. Though not as paranoid as the previous generation, they are numb to politics, which tends to make them conformists to the political ebbs and flows. Also, they did not receive as much formal education as their successors.

Many of this age cohort were placed into the astronautics and aeronautics fields instead of more prestigious engineering fields such as computers, communications, and electronics. Most did not choose to enter into the aerospace field because this was not the most desired field at the time of their graduations. For example, one aerospace engineer who designs spacecraft for the Harbin Institute of Technology (HIT) stated he did not particularly want to do aerospace engineering. However, once placed into the field, he learned to like it. Another communications satellite project manager stated, as a son of intellectuals, he majored in electrical engineering and was selected to work in the aerospace field. He has since come to love his job and strives to do his best. Just as Wei Zhuanfei (魏传锋), a Tiangong-1 space life sciences expert, felt great pressure and humility in his career, the 7th generation does not expect recognition nor will they receive it.[11] In their young lives, they were followers. Now middle aged, they have become indecisive conformists who largely are pessimistic about their futures and long for obsolete values held by previous generations. This age cohort is the last generation in China to have experienced large-scale conflict at the hands of their own government. As they become elders, they likely will be little respected. Very few will become aerospace leaders in the 2030s. More likely, younger and more capable aerospace experts will rise to the forefront to lead China's goals in space.

---

[11] "天宫一号总体设计博士们的故事," 中国航天报, 14 October 2011; available from http://www.spacechina.com/n25/n144/n206/n216/c124983/content.html.

Rise of the Prophet-Engineers

The young generation being groomed by the current leadership in China, largely via the Communist Youth League,[12] is the 8th age cohort of aerospace professionals. The goal of replacing the older professionals with scientists and engineers averaging 35 years old and chief designers in their 40s has largely been accomplished. This generation is becoming the backbone of the aerospace industry[13] and will continue under the new Politburo under President Xi Jinping (习近平). Started during the 11th Five-Year Plan (2006–2010), the goal of inserting the younger generation into aerospace jobs has resulted in most of the current manned spacecraft and rocket developers being around 35 years old. This is an accomplishment greatly touted by the central government. As President Hu stressed during a rally to celebrate the Chang'e-2 lunar mission, the old scientists and engineers served successfully as role models for their younger colleagues, who ingrained them with noble spirits and good work ethics and, thus, the younger generation deserves to be promoted.[14] The average age of the technical staff who worked on the Shenzhou VI human space mission was 32 years old.

The human space program purposefully transferred out experienced technical personnel from the Shenzhou V mission to allow 8th generation professionals to work on Shenzhou VI. This program has partnered with Beihang, Qinghua University, and Northwestern University in the United States to jointly train aerospace graduates.[15] Since 2001, CASC has recruited from 20 technical universities across China and offers competitive scholarships in 31 Chinese universities. CASC claims that of 300 chiefs and deputy chiefs, 60 % are less than 45 years old.[16] In similar fashion, CAS has implemented a training program where graduates spend a year working on their dissertations at one of the CAS institutes under a consolidated graduate program. Around 6,000 students are working on doctorate degrees under this program, thus putting highly qualified and young engineers directly into positions at CAS.[17]

Part of the reason for such successful recruitment of the 8th generation into space leadership positions has been due to the central government's promulgation and implementation of the "liang zong" (两总) management system across the

---

[12] 缪春梅,史吉锋， "以目标为导向的学生干部培养方法研究," 桂林航天工业高等专科学校学报 1:61 (2011) 63–65.

[13] 张晓祺,柳刚, "走向叩问天宇新征程-访中国工程院院士,载人航天工程首任总设计师王永志," 中国军网, 22 December 2010; available from http://chn.chinamil.com.cn/xwpdxw/2010-12/22/content_4356954.htm.

[14] 胡锦涛, "在庆祝探月工程嫦娥二号任务圆满成功大会上的讲话," 新华网, 20 December 2010; available from http://news.xinhuanet.com/politics/2010-12/20/c_12900124.htm.

[15] 孙宏金, "'两总'系统年富力强 青年人才敢挑大梁," 搜狐新闻, 12 October 2005; available from http://news.sohu.com/20051012/n227177715.shtml.

[16] "Talented Personnel;" available from http://www.spacechina.com/n16421/n17221/index.html.

[17] "Education and Training;" available from http://english.cas.cn/ET/200908/t20090826_34258.shtml.

aerospace industry.[18] In 2006, the CCPCC set up this management system in SASTIND. It establishes a commander and chief designer to co-manage each system as a way for an elder leader to train, select, and assess management personnel of the younger generation in support of civil-military integration of space programs.[19] Within the aerospace industry, the management system is broken into five teams responsible for satellites, rockets, launch sites, control systems, and ground systems.[20] Begun under the leadership of Sun Laiyan, CNSA has implemented this system since 2009 to raise management standards and promote professionals with high "political fiber" and loyalty to the CCP.[21]

A lot of hope and pressure has been placed on this 8th generation of aerospace professionals, so much so that they are expected to outdo the achievements of previous generations of aerospace engineers, scientists, and technicians. This is a unique generation that will make the previous generations nervous due to their uncanny confidence in a new advanced China. This is the first generation alive today to never experience civil war, foreign invasion, or mass starvation. They were often born as only-children to intellectual parents and two sets of doting grandparents. They consist of an imbalanced number of young males about to reach marriage age, so they are professionally driven to make good salaries and attract potential wives.

Although most Chinese are familiar with their history, to this generation, Chinese history is simply stories they have heard and studied and not experienced on their own. They largely are privileged middle-class children born into a secure and peaceful environment. Since they have been born into a uniquely peaceful environment, their lack of experience of strife can, at times, distance them from their older colleagues.

The professionals of the 8th generation are generally very innovative and creative. They likely have overseas experience, speak English, and understand China's current and future hierarchical Confucian-based place in the global space community. Uniquely, their place in patron-client relationships extends beyond their chain of command. It also branches out into the business world, across space programs, and international organizations, all of which helps break down stove-pipes. Despite sometimes being called "little emperors," at least in jest, and having a more individualistic approach to their careers, they may well be capable of building aerospace technologies and systems that could, in time, overshadow the Western aerospace industry's capacity for innovation and competition. With the

---

[18] 邝勇and赵本利，"军用关键新材料:国产化及型号替代工作实践，" 航天工业管理 (June 2011) 60-64.

[19] "加大对型号'两总'的管理力度，" available from http://www.gov.cn/ztzl/gfkgw/content_600461.htm.

[20] For an example from the Chang'e-2 management system, see "五大系统两总，" 30 September 2010; available from www.miit.gov.cn/n11293472/n11293877/n13408489/n13416218/13416379.html.

[21] "孙来燕到航天二院考察两总系统履职情况，" 4 November 2009, available from www.cnsa.gov.cn/n615708/n620172/n677078/n751578/169198.html.

high pressure placed on them to succeed by their families and a safe foundation from which to build their careers and families, this generation of aerospace professionals share a unique vision for China's future in space—one that may or may not completely track with the current central government's vision.

As children, many of the 8th generation were ingrained with can-do attitudes and access to newly provided technologies as toys. Members of this age cohort largely made high marks in university and were eager to enter the aerospace field. What once was viewed as a field of study to avoid, aerospace engineering is now deemed potentially lucrative. As professionals, they are well established to make good salaries, gain prestige, and increase their professional opportunities. A good example is Shao Limin (邵立民), who was born in 1979. He is a Northwestern University Ph.D. graduate in aircraft design who chose to enter the aerospace field in 2008 and work on the Shenzhou program.[22]

By mid life, in the 2030s and 2040s, the 8th generation age cohort will likely begin to actualize their ideals in ethics and morality thanks to President Hu's values training of putting political integrity first.[23] They have a clear vision of China's role in the global space community, and they have great potential to accomplish sinification in space endeavors. As adults, they tend to be fiercely nationalistic, which, as mentioned, can be threatening to the CCP. In fact, with high college marks and in-demand skills, they are prone to "job hopping" (跳槽) every 5–8 years, a new phenomenon in China, and are no longer tied to the danwei (单位) system,[24] which also worries the CCP. The Party must deal with how to control a mobile, highly skilled, and needed generation of aerospace professionals. They feel they have "kaituosheye" (broadened their horizons 开拓视野) compared with their predecessors, which could serve to fuel a sense of superiority. As the 8th generation ages, the professionals will continue to be very busy in their careers, continuing to create their vision of China in the world, yet will begin to question their ethics and values as they find themselves isolated from previous generations and potentially disassociated with the rising 9th generation. The 8th generation professionals, who once were the centers of their world, may, in later life, find it difficult to relate to the other generations and, this, in turn, will threaten continuation of the vision of sinification in space they spent their entire lives building.[25,26]

---

[22] "天宫一号总体设计博士们的故事," 中国航天报, 14 October 2011; available from http://www.spacechina.com/n25/n144/n206/n216/c124983/content.html.

[23] "Talented Personnel;" available from http://www.spacechina.com/n16421/n17221/index.html.

[24] Danweis are work units usually associated with SOEs. The danwei can encompass health care, living quarters, and recreation. It also serves as a good method to track employees in terms of travel, family life, and even eating habits.

[25] The 9th generation of aerospace professionals will be briefly discussed in the future scenarios chapter.

[26] One exception to the analysis in the section and one that deserves more attention in future studies is the aerospace professionals who have served in the PLA. Aerospace engineers, scientists, and technicians who have military service tend to transcend age cohort cohesiveness

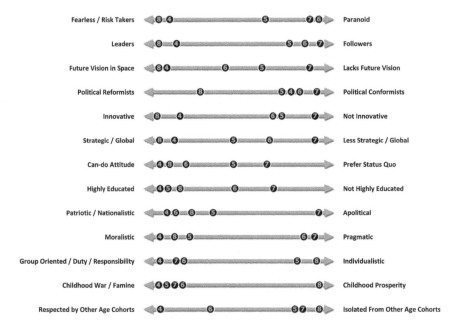

Fig. 5.1 Social and environmental influences on the age cohorts of China's aerospace industry. *Numbers* represent the generation. 1st–3rd generations are not included due to their advanced ages

The above age cohorts of aerospace professionals have varying dimensions of commonalities. There are some that have a number of charismatic leaders and others that have more loyal followers. Some are accepting or while others are more distrustful of central authorities. Some are humble while others are prideful. Some of the cohorts tend to be innovative thinkers and others are more comfortable with more structured thinking. Above is a chart to help visualize how these age cohorts compare to one another (Fig. 5.1).

The purpose of this chapter is to help Westerners understand and communicate more effectively with their Chinese counterparts in space relations. One should not consider the above analysis for more than it is—an overture of who the cohorts are and what motivates their actions. Clearly not everyone in the various cohorts act the same. In short, the purpose of this chapter is to create a general introduction of various cohorts of the leadership personnel in China's aerospace industry. It only is one of several tools to consider.

(Footnote 26 continued)
and instead share military bonds based on an esprit de corps. Also, with the recent push to integrate civil-military space programs, the military professionals in the space programs may be competition to the 8th generation's rise in power.

References

1. Strauss, William, and Neil Howe. 1991. *Generations: The history of America's future, 1584 to 2069*. New York: William Morrow and Company Inc.
2. Lam, Willy. 2010. *Changing of the guard: Beijing grooms sixth-generation cadres for 2020s*. Washington: The Jamestown Foundation.

Chapter 6
Chinese Space Ethics and Decision-Making

Introduction

Chinese ethics exist within and influence the aerospace leadership's decision-making processes [1]. Understanding how Chinese ethics play into these processes will give a better understanding of how and why Chinese aerospace leaders make decisions. Chinese ethics are based on an internal humanistic value system while moral principles based on divine authority are found in Western Judeo-Christian traditions. The Chinese philosophies of Confucianism, Buddhism, Daoism, and Legalism have influenced modern Chinese space ethics to varying degrees. This section seeks to provide a better understanding of Chinese ethics that play a critical, albeit oftentimes subconscious, role in leadership decision-making within the aerospace industry.

Just as traditional American values, such as liberty and rugged individualism, have remained prominent in American society and culture, Chinese values have played significant roles in political design over the years. Chinese scholars, aware of the ethical significance tied to importing Western science and technology, carefully study Western ethics in an attempt to preserve deeply rooted Chinese ethics, especially Confucian values, and curb importation of tacit influences from Western technologies. Chinese leadership aims to regulate aerospace engineers' and scientists' morals and behaviors via dissemination of ethics through official promulgation processes and formal education [2]. For example, in 2010, CASC utilized the power of promulgation and published the first report on social responsibility to inform and promote specific Chinese-based ethics within the aerospace industry.[1] CASC leaders believe moral education is also an important part of promoting ethics within the aerospace industry. CASC teaches social ethics in in-house training programs for "harmonious" team building. CASC's website even has a "social responsibility" link prominently displayed on its homepage.

---

[1] "Social Responsibility Report: 2010年度社会责任报告," 中国航天科技集团公司; available from http://www.spacechina.com/n25/n142/n158/n4604/c146209/part/146221.pdf.

S. Solomone, *China's Strategy in Space*, SpringerBriefs in Space Development, DOI: 10.1007/978-1-4614-6690-1_6, © Stacey Solomone 2013

Ethics in the Aerospace Industry

Harmony (和) and benevolence (仁) are central ethical concepts influencing and permeating the aerospace industry, and which play roles in internal decision-making processes. The pursuit of harmony takes place within a person through self-cultivation. Harmony also is desired between people via their relationships or guanxi (关系)[2] and between nation-states via peaceful coexistence. Based on both Confucian and Buddhist principles, harmony is not static, but instead naturally occurs in a state of constant flux. Therefore, to attain harmonious flux within a person, guanxi, or nation-states is to attain an acceptable and natural state of peace for the Chinese.

Cultivating benevolence within a person includes adopting moral behaviors of an individual within parameters expressed by Confucius' student Mencius. According to Mencius, a person should not be too passionate, but should be able to regulate his behaviors and conform to social norms. To attain benevolence, one must investigate things and pursue knowledge [2]. Therefore, aerospace professionals oftentimes do not share the same wanderlust or inspiration to boldly go where no one has gone before as many Western aerospace professionals have. Instead, what is seen is an outward appearance of a pragmatic and inquisitive exploration of space for practical applications or an investigation based on the desire to solve scientific or technological puzzles.

As the 2010 CASC Social Responsibility Report reads, a core value is "harmonious space" (和谐航天) from which aerospace employees can form respect, a good working atmosphere, and camaraderie between employees by granting them pension programs, workers' compensation, maternity leave, and medical insurance.[3] At the nation-state level, as written in the 2011 Space White Paper, principles of equality, mutual benefit, and peaceful utilization of outer space are extolled, as is the development of space for the common wealth of mankind. These words sometimes appear insincere to the Western eye, but when taken within the context of harmonious flux and cultivation of benevolence, they better illuminate China's fundamental approach to space endeavors. The goal is to build peace, prosperity, and stability, all of which factor into the desired end state of harmonious flux. Because harmony is not a fixed state within an individual or between people or nation-states, the interdependent relationships must be iterative and adjust to continuous changes. Therefore, the aerospace leadership decision-making process tends to be iterative and adjusted on a continuous basis. There is no finality to decision-making.

What of man's relationship with nature, specifically with space? According to Chinese ethics, human life is determined by the cosmos. Science allows discovery and understanding of the cosmos, which will lead to a better understanding of life, and, thus, humans will be able to create better norms for how to live [3].

---

[2] Guanxi, or relationships, are long term ties based on mutual obligation and reciprocity.

[3] See footnote 1.

Specifically within Daoist beliefs, humans are a part of the whole cosmos. Therefore, both life and death should be celebrated because both are natural states of coexistence in the cosmos. Some may wrongly believe that the Chinese do not value life. Indeed they do. They also value death. Death is a part of life and a part of the cosmos [4]. This helps the Chinese maintain resilience in the face of death because of one's identification with the whole cosmos. Daoism supports this ethical belief that one should not fear hardships, death, or misfortune. Indeed, a common phrase in Chinese is to "eat the bitterness" (吃醋) as a way to accept and internalize negative emotions.

Since man is regarded as a part of the cosmos, then the relationship between humans and space is one of mutual benefit. Harmonious flux, as discussed above, is the desired end state. Therefore, instead of a conquer-and-prevail attitude toward the space environment, which may pit nation-states against each other, the Chinese seek to not *own* space but to fully *utilize* space. Thus, Westerners may read and hear references to the "peaceful utilization of outer space" in China's official texts and during international exchanges on space matters. According to Chinese philosophy, humans should go so far as to adapt to their environment, including space, since humans and the environment are two parts of the same cosmos.

So, this leaves a very interesting conundrum. Why did the Chinese conduct the 2007 anti-satellite (ASAT) test, leaving an estimated 35,000 pieces of orbital debris and thousands of trackable objects (i.e., greater than 10 cm in size) in low Earth orbit and, thus polluting the space environment for perhaps centuries? [5]. The answer is not in the test itself but rather in the Chinese government's and domestic scientific community's responses and successive activities since the test. The Chinese are fallible; they do make mistakes and miscalculations. Judging by subsequent words and actions post-ASAT test, the Chinese government and scientific aerospace community have boisterously proclaimed how bad orbital debris is for the space environment. It is unlikely we will ever again witness another test of that nature from the Chinese. It goes against their space ethics.

Space Ethics in Decision-Making

Space ethics permeate the aerospace community and influence their decision-making processes. Chinese space ethics, as very briefly described above, manifest in dialectic thinking and dichotomous approaches to decision-making.

First, dialectic thinking embraces many complex elements that are different from one another, mutually dependent for their existence and continuously evolving. It is a different concept than the black and white approach often found in Western cultures. To illustrate, if one were to ask if the proverbial glass is half empty or half full, a Westerner likely would chose one or the other based on his positive or negative outlook. For the dialectic thinker, the answer simply is "yes." This type of thinker can see that, indeed, the glass is both half empty and half full at the same time without having a conflict of two opposing states existing at the same time and place.

Dialectic thinking is used in decision-making by taking a holistic perspective that examines all distinctions and linkages between elements. In addition, it helps to balance short-term goals with long-term goals. The Chinese have a long-term view of their goals in space that may be due to this type of decision-making process. Dialectic thinking also incorporates the dynamics of change and contradictions by scoping them within decision-making processes via social responsibility, management techniques, and creation of corporate culture. Dialectic thinking allows for consideration of all technologies so long as they have some evidence of future viability. Thus, this lends credence to the point that the Chinese may explore space-related R&D programs simply based on the notion of scientific curiosity and intellectual learning. Dialectic thinking is iterative; it embraces criticism and self-criticism, and takes all theories into consideration [6].

Second, in addition to dialectic thinking, dichotomous approaches to decision-making, also seen in the glass illustration above, embrace the thought process that two seemingly opposite, or mutually exclusive, choices, can coexist. Oftentimes, Chinese aerospace interlocutors have been charged by their Western counterparts as "speaking out both sides of their mouths." Ambiguous messages abound, and seemingly little progress is made in the context of oftentimes sensitive space-related relationships between nation-states. For the Westerner, if the incompatibilities are not removed, then an air of tension may result [4]. But, if the Chinese concept of dichotomy toward space relations is taken into consideration, perhaps an understanding can occur that incompatibilities can coexist within agreements. These dichotomies, from a Daoist perspective, have roots in the concept of yin and yang (阴阳), where two components are opposite yet must coexist in a harmonious state [4]. This is why on one hand, one can witness a rather humble Chinese aerospace leader, yet also see a prideful and unbending negotiator. As was discussed in the age cohorts chapter, character traits of the aerospace professional can reflect a dichotomous nature at times. Opposite qualities can simultaneously exist. It is also why we can see how a foreign space technology can be viewed by Chinese as an indigenous technology (albeit recombinative). Finally, it helps put into context the problem of plagiarism within the Chinese space science community.

An issue that has plagued Sino-U.S. space relations since the January 1999 Report of the Select Committee on U.S. National Security and Military/Commercial Concerns with the People's Republic of China, known as the Cox Report,[4] is that of transparency into Chinese space endeavors. According to Sun Tze (孙子),

---

[4] The Cox Report was intended to restrict American transfer of technologies that China also could use in military applications. What was intended to restrict advanced aerospace technologies from entering China, may arguably have been the catalyst for their aggressive pursuit to indigenously develop (via recombinative and independent innovation means) aerospace technologies.

it is more important to see an enemy's intentions than to see the capabilities. It is a sign of weakness to show one's capabilities. Using secrecy to hide one's capabilities is a means to retain harmony and balance. Revealing capabilities would create an imbalance. It seems counterintuitive to a Western strategist, but not from a Chinese perspective.

For example, the Confucian concept of filial piety stresses that an obedient son should do whatever it takes to honor his father. Thus, in some cases, it is better to conceal the truth from one's father, even to the extent of lying, if it is necessary to protect and honor him and maintain filial piety. In the case of national security issues in space, one can see where this concept could come into play. From the Chinese perspective, mutual dependence among actors in the global space community maintains balance. Because the Chinese clearly stated their intentions in space in the Space White Papers, such as peaceful utilization of outer space for the common good of mankind, then there is no need to reveal their hardware capabilities.

The same holds true in internal space programs. In 2011, CASC issued a manual on the "Culture of Secrecy"[5] within the domestic aerospace community. By 2015, CASC plans to have in place a new system of secrecy that will meet the challenges from integrating civil and military space programs. The goal of the manual is to put forth and begin practicing the concept of embracing secrecy to shape behavioral norms. Just as with the Confucian concept of filial piety, one must protect the Chinese motherland to the extent of concealing truths from one another, all to maintain harmony within the aerospace industry. However, there are apparent weaknesses in this approach to security within the domestic aerospace industry in terms of corruption, stovepiping, and redundancies between programs.

Methods of Decision-Making

Three methods are vying for primacy within the Chinese aerospace industry—rule by reputation, rule by consensus, and rule by law. These competing decision-making processes can lead to apparent strife among the aerospace leadership, and, each in its own unique way, can lead to the ethical dilemma of inaction.

First, in recent Chinese history, rule by reputation was key to success, as seen in the leadership of Mao Zedong and Deng Xiaoping. Rule by reputation is based on the Confucian model of ethical nobles ruling by example and demonstrating how one ought to conduct oneself when in power via moral exhortation. Rule by reputation supports the notion that if an emperor does not act in the best interest of the people, then the people have a duty to inform the leadership of the right course of action. However, as illustrated by the concept of filial piety, sometimes it is

---

[5] "中国航天科技集团公司《保密文化手册》解读," 中国航天报, 16 September 2011; available from http://www.spacechina.com/n25/n144/n206/n214/c67226/content.html.

better to conceal the truth to show respect for the father (here, the leadership), which may lead to a breakdown in command and control. Two opposing ethical approaches that coexist may lead disgruntled aerospace engineers to choose to "eat the bitterness" rather than complain about poor work conditions, for example, even though it may affect quality control. The opposing ethics here can lead to a state of inaction.

Second, decision-making under rule by consensus is a slow, methodical, and pragmatic process. Aerospace leaders work very slowly to carefully analyze challenges. Consensus building takes time and is rooted in the ethical concept of the group's good over that of the individual. Consensus building supports the ethics of inclusiveness and acceptance, which is also considered during the dialectic thinking process. Thus, decision-making under the rule by consensus method is an iterative process. What, at first, appears to be a detriment to the need for speedy decisions when in a time of crisis has, in fact, worked well for Chinese aerospace leaders.

For example, the long, arduous process of fighting ProtoStar, Ltd. within the legal confines of the ITU, resulted in slowly draining money from ProtoStar, Ltd., which ultimately ended up declaring bankruptcy. The process likely gave Chinese leadership the time needed to build consensus with those affected by the alleged interference from the Protostar-1 satellite. Arguably, it may be that the nature of space issues does not require swift decision-making at the higher echelons, thus resulting in a very effective decision-making process within this industry. For the sake of argument, if rapid attention were needed, say, in the case of orbital conjunctions, where it may be necessary to move an active satellite to avoid a collision with orbital debris, then this type of scenario should already have in place a process at the lower echelons and should not require the type of leadership decision-making cycle discussed here. If, however, Chinese aerospace leaders were constrained by the need for rapid decisions, then it may be detrimental to their desired outcome.

Without a charismatic leader within the domestic space community, Chinese decision makers have had to rely on consensus building. Rule by consensus has roots in Daoism which stresses what leaders should, or should not, do to maintain a balanced status quo (or harmonious flux). Also, Daoism encourages the study of history to look for the right decision. Looking to history for guidance can restrict innovative thinking by leaders. For example, the DFH-3 satellite bus, which was initially built with German assistance and has been very successful, illustrates how relying on recombinative innovation and looking to what already existed can restrict truly independent innovations for future satellite buses. Maintaining the status quo and looking to history for guidance are two concepts in Daoism that may lead to inaction by leadership in the aerospace industry.

Third, partly due to pressures from the global space community, Chinese aerospace leaders are using the rule by law method in decision-making. With roots in the ancient Chinese philosophy of Legalism, decision-making that incorporates rule by law is based on the concept that institutional structure is needed to influence people to behave in certain ways via legal coercion [4]. Legalism seeks to punish those who not only fail to perform but also those who do more than they are assigned to do. Legalism is also used as a form of coercion to pressure actions or inactions within the aerospace industry. This can lead to mediocrity and kill innovation by warning those aerospace engineers who come up with new and innovative ideas that their ideas threaten norms, established corporate space culture, and current direction of space programs, thus, resulting in another form of inaction.

The above cited three methods of decision-making, each in its own unique way, can lead to the ethical dilemma of inaction. All three forms continue to vie for control over decision-making within the aerospace community, which is consistent with the concept of continuous and harmonious flux. It is important to understand, or at least to be aware of, the nuanced differences between Chinese and Western space cultures and ethics when discussing space matters. Understanding the roles of Chinese space ethics and aerospace leadership decision-making processes can assist Westerners when engaging with China on space issues.

References

1. Joseph, Needham. 2004. China's immanent ethic. In *Science and civilisation*, Vol 7, 84–85. Cambridge: Cambridge University Press.
2. Guo Xiaoping and Song Enrong 2004. A case study of chinese core values of peace and harmony. In *Teaching Asia-Pacific core values of peace and harmony: A sourcebook for teachers*, eds. Zhou Nanzhao and Bob Teasdale. Bangkok: UNESCO. Available from http://www.unicef.org/violencestudy/french/pdf/Teaching%20Asia-Pacific%20core%20values.pdf.
3. Zhou Changzhong. 1998. Ethical reflections on western science and technology in the philosophy of modern China. In *The humanization of technology and chinese culture,* eds. Tomonobu Imamichi, Wang Miaoyang, and Liu Fangtong (Washington: The Council for Research in Values and Philosophy). Available from http://www.crvp.org/book/Series03/III-11/chapter_vi.htm.
4. David Wong. 2008. Chinese ethics. In *Stanford encyclopedia of philosophy*, ed. Edward Zalta, 10 Jan 2008. Available from http://plato.stanford/edu/entries/ethics-chinese.
5. Kelso, T.S. 2007. Analysis of the 2007 Chinese ASAT test and the impact of its debris on the space environment, (paper from AMOS Conference, Maui), 321–330. Available from http://celestrak.com/publications/AMOS/2007/AMOS-2007.pdf.
6. Gao, Xudong, Yu, J., and Li, M. 2010. Developing effective strategies to address complex challenges: evidence from local high-tech firms in China. *International Journal of Technology Management* 51(2–4):324, 333.

Chapter 7
The Aerospace Industry and Chinese Society

Effects of Chinese Culture on the Domestic Aerospace Industry

A unique Chinese space culture has emerged within the domestic aerospace industry. The space culture has been shaped by socio-historical trends, which is reflected in different stages of China's space endeavors [1]. China moved from a newborn state focused on national identity and pride to one focused on political correctness as defined by Mao Zedong and the CCP. Next, China transformed into a nation fixated on economic endeavors, and then one focused on national identity and cultural heritage. Currently, China's long-term future vision is directed toward China's place in the world. By seeing China's space culture emerge over time through these stages of recent history, it is then possible to address how cultural influences impact the aerospace industry now and in the future (Table 7.1).

The 1950s was a flourishing time for China's aerospace industry. Built during a thriving time in China's evolution, the aerospace industry carried a positive vision of China's future and its place in space. With the Red Army as the foundation for establishing the PRC, the dream to pursue space activities was tied to military power. But the political currents rapidly shifted as the period of 1960–1965 was marked by extremes in political power held by Mao Zedong. The fledgling space program flourished with guidance from the Soviet Union, and great strides were made in developing the rocket program. The CZ rocket family was named after a very famous event in China's politico-military history. In 1934, the Red Army, consisting of Mao Zedong, Zhou Enlai, Deng Xiaoping, Zhu De (朱德), Peng Dehuai (彭德怀), and 86,000 followers, was being pursued by over 300,000 Guomingdang (国民党) military forces. Forced to flee on 16 October 1934, the army left Jiangxi Province and endured terrible hardships, including multiple battles with the Guomingdang, freezing weather during the mountain crossings, and exhaustion. Just over one year later, on 20 October 1935, the Red Army arrived in Shaanxi Province. Only 4,000 people survived; however, the Red Army had accomplished the great feat of covering 6,000 miles to finally arrive to safety.

S. Solomone, *China's Strategy in Space*, SpringerBriefs in Space Development, DOI: 10.1007/978-1-4614-6690-1_7, © Stacey Solomone 2013

Table 7.1 Trends in Chinese space culture

Years	Cultural trend
1949–1960s	National identity and pride
1966–1976	Political correctness
1977–1989	Economic endeavors
1990–2005	National identity and cultural heritage
2005 to Present	Future-oriented vision

Because the Chinese people hold such solemnity and remembrance for the people who endured the "Long March," it is with great pride that China named its rocket family after the event, which embodies the culture of politics of this time.

Unfortunately, the Chinese sense of national identity and pride evolved into a sense of extremism as Maoism reached its peak. The Cultural Revolution was a very dark period in the PRC's short history. But, space endeavors survived despite the chaotic political times when scientists and engineers were often persecuted if not protected by the patronage of Zhou Enlai or Nie Rongzhen. The country's educational system was shut down, and foreign technical assistance was banned. China accomplished a great feat in China's space history on 24 April 1970, when the first DFH satellite was launched into orbit. The name has its roots with a very popular song of the Cultural Revolution, "The East is Red," which was transmitted by the satellite back to China for all the people to hear. The lyrics demonstrate how central Mao Zedong and the CCP were to the people's daily lives during this time. This song exemplifies how the culture of politics permeated everything during the Cultural Revolution and how leaders were able to relate the successes of the space programs to the Chinese people.

"The East is red, the sun has risen, China has made a Mao Zedong. He creates fortune for the people, Hu er hai yue, he's the savior of them all! Chairman Mao loves the people, He is our guiding leader, For developing a New China, Hu er hai yue, leading towards progress! The Communist Party is like the sun, It brightens up everything it shines. Where there is the Communist Party, Hu er hai yue, people are liberated!"

Once the Cultural Revolution ended, China took on a more pragmatic approach to politics. Deng Xiaoping's Four Modernizations allowed the Chinese to explore a limited market economy. As China opened up to the world, businesses were newly created and joint ventures were aggressively sought. In the 1980s, Chinese culture was greatly influenced by this new economy-based society. People in the aerospace industry began to see space as an opportunity for exploiting resources and gaining economic advantages. The 1988 CBERS joint venture between China and Brazil is a good example of the cultural acceptance of China toward foreign business ventures. The Chinese name for CBERS, "Resources," (资源) exemplifies a sense of practicality in China's space culture during this timeframe. During the 1980s, growing economic influences made a great impression on Chinese space culture as the Chinese people were obsessed with making money, creating businesses, and growing the national economy.

It was not until the 1990s that Chinese space culture began to change again. In April 1992, China's second effort to create a human space program was launched after such initiatives were first proposed in the 1960s. In March 1995, China and Russia signed an agreement for technological assistance that included training Chinese taikonauts. The first Shenzhou unmanned capsule launch was scheduled for 1 October 1999, on the 50th anniversary of the founding of the PRC, although ultimately the launch was postponed until 19 November. The name Shenzhou, or "Divine Vessel," signifies recognition of China's ancient culture. The CZ rocket that carried the Shenzhou capsule into orbit was renamed Shenjian ("Spiritual Arrow" 神箭). President Jiang Zemin handwrote the characters that were transposed onto the side of the rocket body. The Shenjian name has origins from the word huojian (火箭), which is the name of an ancient Chinese rocket that was made of a long arrow with a small bamboo container of gunpowder attached to the back of the arrow and lit for launch. The naming of the capsule and rocket for China's human space program indicated a cultural sensitivity to China's ancient history that had been banned during the Cultural Revolution. Also, it indicated a re-emergence of Chinese national pride, albeit histo-cultural pride, similar to what was once so prevalent in the early 1950s.

By 2005, China's space culture changed again from one focused on ancient history to its current manifestation focused on the future. New aerospace endeavors, such as near-space dirigibles and deep space exploration platforms to support R&D in materials sciences, energy conservation, and fluid physics, illustrate China's push into long-term future-oriented aerospace endeavors. China is also interested in human space exploration of the Moon by 2030 and establishment of a lunar-based astronomical observatory by 2045. By 2050, China seeks to accomplish the world's first human landing on Mars [2]. China's space culture today is a complex mix of ancient historical pride, pragmatism, and long-term future visions. In addition, with the rise of China's aerospace prophet-engineers, it would not be surprising to see the emerging space culture envelop Western space culture practices.

By taking a deeper look at cultural structures within space activities, one can explore how these structures cause and influence activities in the aerospace industry. Because differences in language, history, and culture abound between China and the Western world, it is important to be sensitive to China's space culture in order to understand how aerospace activities may act as a vehicle by which Chinese people view themselves and gain better understanding of how Chinese space culture has affected China's concept of its place in relation to the rest of the global space community. For example, although initially there seems to be little room for myth to take place within the scientific realms of the space programs, myth, indeed, has seeped in.

The myth of Pan Gu (盘古) is the story of the beginning of creation in which everything was in a state of chaos. Then, Pan Gu was born, in a cramped space. So he began to stretch, and as he stretched, he separated Earth from the heavens. Every day, the heavens rose 10 feet higher, Earth grew 10 feet thicker, and Pan Gu grew 10 feet taller. This took place over 8,000 years until Pan Gu died. His left eye became the Moon, his right eye became the Sun, and Pan Gu's hair became the stars. On Earth, Pan Gu's breath became the wind and clouds, his voice became thunder, and the lice and scabies on Pan Gu's body became men. Just as Pan Gu took a long time to create the universe, the Chinese aerospace industry has its eyes set on a long future in space. From the Pan Gu creation myth, one also can appreciate how vastly different man's place in the universe is compared to that of the Judeo-Christian creation story in Genesis that describes how Earth was made in six days and man was made in the image of God.

Another example of myth in China's space culture is the Chinese name for CLEP's lunar orbiters, Chang'e (嫦娥). The name comes from an ancient legend of a beautiful woman who resides in the Moon. This woman, Chang'e, secretly drank an immortality elixir. As she became immortal, Chang'e floated up to the Moon, where she forever resides. This legend is a central story in Chinese culture. As the Chinese aerospace professionals reach toward the Moon and eventually toward Mars, they are bringing Chinese space culture and practices with them, which will permeate all aspects of space activities, as it is impossible to separate technology from culture. Therefore, it is important to know about Chinese space culture to appreciate and communicate with the Chinese about their visions for space activities.

Effects of the Aerospace Industry on Chinese Society

Many satellite applications have touched the lives of the Chinese people by facilitating rescue and relief efforts during natural disasters, helping feed the people, and making it more convenient to live in a modern society. To illustrate how satellite applications have been used to help save people's lives during natural disasters, on 12 May 2008, an earthquake measuring 8.0 on the Richter scale hit Sichuan Province and resulted in 68,000 deaths. During the rescue and relief efforts, 15 satellites with nine different applications were used to provide emergency services. In circumstances where communication networks were destroyed and roads were badly damaged, China's satellite navigation constellation provided vital information to the first responders and aided them in appropriate rescue measures. First responders used the Beidou navigation constellation to send short text messages in areas where all other communication modes were destroyed. More than 1,000 Beidou terminals were engaged in the rescue and relief efforts to establish communications between different rescue teams and between rescue teams and the Beijing rescue headquarters.

As Vice Director of the China Satellite Navigation Engineering Center Ran Chengqi (冉承其) stated, "the Beidou system is not as far removed from people's daily lives as many believe…" [3]. With bilateral and multilateral humanitarian assistance and disaster relief (HADR) efforts being sought with China, it is important to bear in mind that domestic disaster relief must be conducted internally by the CCP and PLA so as not to lose face. In ancient China, it was the responsibility of the emperor to protect the people from bad weather and failed crops. When a foreign military offers HADR assistance, it may cause CCP and PLA officials to feel as if they are viewed as not being able to fulfill an ancient responsibility to save the Chinese people during natural disasters.

In addition to rescue and relief efforts, the CCP also must ensure the people have food, which helps retain legitimacy. Liu Luxiang (刘录祥), director of the Space Technology Breeding Center, stated, "Breeding seeds in space is expected to become a strong driving force behind Chinese agriculture in the 21st century since it can bring about high-yield and high-quality crops that are hard for ordinary breeding methods to obtain" [4]. Space breeding refers to a technique using the high vacuum, low gravity, and strong radiation space environment to mutate the seeds. Upon returning to Earth in a recoverable satellite, mutated seeds are selected and planted to breed strains with higher quality and yield, and which may be resistant to disease and early maturing.

China is believed to be the only country seeking to apply the process to large-scale terrestrial agricultural farming. According to Liu, since 1986, China has launched eight recoverable spacecraft with over 70 varieties of crop seeds on board. These seeds include rice, cotton, vegetables, and fruit. One mission that carried seeds into space was the maiden flight of the Shenzhou manned capsule. Carried onboard were seeds from tomatoes, watermelons, radishes, green peppers, corn, barley, wheat, and over 30 different types of herbal medicine. According to Chinese scientists, space seeds from rice when compared to ordinary strains had 10 to 15 % more yield per unit. Tomatoes grew as heavy as 800 grams each after seven and a half years of cultivation. Bell peppers grown from the space seeds had 25 % higher yield and 20 to 25 % more vitamin content than ordinary strains.

Beyond the fundamental satellite applications to help save and feed people, satellite applications are responsible for creating many life conveniences for the Chinese. For example, China had been a largely cash-based society until the 2008 Olympics helped popularize automated teller machines (ATMs) and credit card usage in Beijing. In the run up to the Olympics, China added credit card terminals at shops and restaurants to ease payments by credit card users. Even aerospace professionals now receive business credit cards bearing pictures of the CASC logo, Shenzhou capsule, and CZ rocket. An emerging Internet-based banking system in China using very small aperture terminals (VSATs) makes it possible for someone to deposit or withdraw money from any of the bank ATMs. There are now over 1.5 billion credit cards in use in China. Bank card transactions account for about 25 % of the country's total retail sales. Also, as more credit cards are adopted, e-commerce will grow, which will increase the potential to monitor citizens' transactions. Use of VSATs for credit and debit cards is making life easier for the Chinese people, but it will also allow increased monitoring of financial transactions.[1]

---

[1] "Inter-Bank Bank Card Transactions Soar on New Year's Holiday," China Union Pay; available from http://en.unionpay.com/news/newsroom/file_4259733.html.

Fig. 7.1 Graffiti of Yang Liwei on a wall in Beijing.[2] (Image provided by the author.)

Effects of the aerospace industry on Chinese society is an important aspect of space efforts for the Chinese central government. The CCP has very effectively used Chinese culture as a tool to give the people a "space identity," which, in turn, legitimizes the government's expensive space endeavors. Yang Liwei's flight proved to be the main event that paved the way for the Chinese people to think of themselves once again as part of a true space-faring nation since the historic 1970 launch of the DFH-1 satellite. The central government's efforts to publicize these events has allowed for all Chinese to share in the successes of the human space flights. Even the poor Chinese farmer can be proud of China's human space program despite not having much impact on his life in rural society. The popular culture of China's space activities can be seen in many ways, such as in the graffiti present in the graphic above (see Figure 7.1).

A large portion of space commercialization has made life easier for the Chinese people. To date, roughly 400 space technologies have been applied to the education, health, and mining fields.[3] In December 2011, China began providing location and navigation services to domestic customers through the Beidou system. In addition, businesses that do not directly support the space programs have successfully marketed products using images of taikonauts, rockets, and satellites. Because of the strong pride and connection the Chinese people feel for the successes of the

---

[2] Author's personal picture.

[3] "China's Aerospace Investment Benefits other Areas: Official;" available from http://news.xinhuanet.com/english/china/2012-06/29/c-131684683.htm.

Fig. 7.2 Chocolate-flavored candy for sale. [4] (Image provided by the author.)

human space program, this marketing tactic greatly appeals to the rural population who can afford, for example, a candy bar or chewing gum. The famous all-female rock music group S-H-E exploited this marketing tactic in 2007 by using an image of the three singers in extra vehicular activity (EVA) suits to sell domestically produced chewing gum. Other images of models dressed in EVA suits have been seen in Henan on milk containers. And even images of the U.S. space shuttle have appeared on large billboards in Beijing to sell wrist watches. Marketing of human spaceflight images has been very effective to help turn out profits for businesses that have no direct connection to actual space endeavors. This marketing tactic ebbs and flows with each successful human spaceflight, resulting in an inundation of images post-launch and then those images fading away until a couple of years later, when the next human spaceflight takes place (Fig. 7.2).

As a form of entertainment, space-related themes have diffused into society. For example, not only has foreign space science fiction gained popularity, such as the *Star Wars* trilogy, but a new collection of domestic space science fiction has blossomed in recent years, taking on increasingly complex social and cultural themes placed in futuristic space environments. Also, a plethora of satellite broadcasts of variety shows has gained popularity in the countryside to the extent that the central government had to place strict limitations on broadcasting entertainment that they

---

[4] Author's personal picture.

deemed low taste.[5] In October 2011, 75 % of the broadcasts were considered "excessive entertainment" and mostly consisted of dating and talent shows.[6]

The rest of this chapter takes a closer look at the causes and effects of Chinese space science fiction on the aerospace industry and Chinese society. China's space science fiction foments imagination and creativity—necessary elements for China's efforts of independent innovation in the aerospace industry. Space science fiction stories reflect the cultural and social dynamics of the push for technological innovation and China's future in space. During a discussion with the author, famous science fiction writer Han Song (韩松) stated, "The unconscious descriptions of ethical problems in Chinese science fiction probably hint at a future reality in the country's space program" [5].

China's first exposure to science fiction began with the works of Lu Xun (鲁迅) (1881–1936), who is called the Father of Chinese Science Fiction. In 1903, Lu Xun translated Jules Verne's famous work *From the Earth to the Moon*. In the preface, Lu Xun lamented that "science fiction is as rare as unicorn horns, which shows in a way the intellectual poverty of our time. To fill in the gap in the translation circles and encourage the Chinese people to make concerted efforts, it is imperative to start with science fiction [6]." Lu's translation of Jules Verne's novel and his own preface were posted on the China Lunar Exploration Program (CLEP) website (http://www.clep.org.cn), which has since been shut down. One year later, in 1904, Xu Nianci (徐念慈) wrote *The New Story of Mr. Fairytale* in which a gust of wind splits the hero's body and spirit. The spirit travels up into the Solar System and then back into the center of Earth. This is credited as being China's first indigenous space science fiction story (Fig. 7.3).

After the founding of the PRC, with a fledgling aerospace industry being developed in the 1950s, space science fiction writers began to emerge. In 1954, Zheng Wenguang (郑文光), an astronomer and space science fiction writer, wrote *From the Earth to Mars*. Similar to reactions by Americans after hearing H. G. Wells' *War of the Worlds*, the Chinese people were so enthralled with Zheng's story that thousands of people rushed to observatories around China to get a closer look at Mars through telescopes.

Science fiction in the 1950s was greatly influenced by the emerging aerospace industry. Stories during this time reflected the naïve illusion that science and technology were the sole ingredients needed in Chinese society to create a positive and hopeful future.[7]

---

[5] "China Limits Entertainment Programs on Satellite TV," 新华, 25 October 2011; available from http://www.china.org.cn/china/2011-10/25/content_23724679.htm.

[6] Ben Blanchard, "China Claims Success in Curbing Racy Entertainment," *Reuters*, 4 January 2012; available from http://tv.yahoo.com/news/china-claims-success-curbing-racy-entertainment-133901427.html.

[7] 刘慈欣,"消失的溪流:八十年代的中国科幻," 7 February 2011; available from http://www.kehuan.net.cn/article/8.html.

Fig. 7.3 CLEP logo

中国探月
C L E P

Despite aerospace endeavors surviving the Cultural Revolution, space science fiction did not. It was not until the 1980s that writers began once again to pick up the pen. However, an evolution in writing style was apparent. If, as Han Song stated, space science fiction reflects realities in the aerospace industry, then a change certainly had taken place. Liu Cixin (刘慈欣), an electrical engineer and science fiction writer, argues that true Chinese science fiction did not really take form until the 1980s, when novel ideas with Chinese characteristics were being created. It was then that China began to cultivate its unique style of space science fiction writings. Stories began to reflect the notion that new technologies could improve morality. Technologies were once again viewed as the way to save society, which is a reflection of Deng Xiaoping's policies during the time. Most notably, Chinese space science fiction began to merge Chinese history and myth into new stories.

With a new vision of humanity's place in the universe, the Chinese people moved from being the lice and scabies off of Pan Gu's skin to heroes who explored space and discovered the mysteries of the universe. Chinese space science fiction began to reflect personal development, morality, and culture. For example, Zheng Wenguang's story entitled *Flying toward Sagittarius* (飞向人马座) is one of sadness, loyalty, and human growth during a long trip in outer space.[8] Space science fiction also served a political function that promoted the vision of China's

---

[8] To read *Flying Sagittarius*, visit http://www.kehuan.net.cn/book/feixiangrenmazuo.html.

evolution into a prosperous and independent country.[9] *The Woman Blowing Smoke Rings* (吐烟圈的女人), published in the early 1980s by an unknown author, tells the story of how a city overcame the problem of air pollution by releasing factory exhaust straight into space. Again, technology serves as the savior of society, reflecting Deng's modernization policies.[10]

Today, Chinese space science fiction continues to be influenced by domestic aerospace endeavors, especially the human space program. Over 50 % of Chinese space science fiction involves space exploration and travel.[11] As space science fiction protagonists traverse the universe in self-discovery, Chinese space science fiction continues to evolve today. China is moving toward an increasingly advanced technological and prosperous economic society, and the Chinese people are becoming more confident in their traditional culture and values, which likely will evolve into an amalgamation of modern Chinese and Western morals, according to Liu Cixin.[12] Dr. Wu Yan (吴岩), a professor of science fiction at Beijing Normal University, echoes the sentiment that technology can enhance morality in China's future.

However, China's new space science fiction themes have been taking on darker topics and oftentimes describe a futuristic dystopia. For example, Liu Wenyang's (柳文扬) *Wake* (醒来) is a story of war between worlds and mind control over soldiers with no happy ending.[13] William Gibson's 1984 novel *Neuromancer* is by far the most famous Chinese-translated foreign dystopian science fiction, which introduced the cyberpunk world to Chinese society. Even though Liu Cixin himself is famous for his *Three Bodies* (三体) dystopian trilogy,[14] he argues Chinese space science fiction writers have a responsibility to create for the reader hope for the future, optimism, and a society based on scientific accomplishments. This sentiment supports the notion that space science fiction writers should be used as tools by the aerospace industry and central government to promote popularity for its space endeavors instead of allowing freedom of speech for Chinese writers.

Most recently, the Chinese central government banned foreign science fiction involving alternative futures and time travel,[15] such as the famous 1985 movie *Back to the Future* and the 1984 movie *Terminator*, due to the potential for Chinese

---

[9] Wu Yan, "Chinese Space Exploration: Science Fiction—An Overview," (presentation at the International Space University, at Beihang University, Beijing, in July 2007) powerpoint presentation.

[10] 刘慈欣, "消失的溪流:八十年代的中国科幻," 7 February 2011; available from http://www.kehuan.net.cn/article/8.html.

[11] See foot note 10.

[12] "Next 60 Years to Witness a New Traditional China," 新华, 29 September 2009; available from http://www.china.org.cn/features/60years/2009-09/29/content_18627776.htm.

[13] To read "Wake," visit http://www.kehuan.net.cn/book/xinglai.html.

[14] To read Liu Cixin's trilogy, visit http://www.kehuan.net.cn/book/santi.html, http://www.kehuan.net.cn/book/heiansenlin.html, http://www.kehuan.net.cn/book/sishenyongsheng.html.

[15] "广电总局关于 2011 年 3 月全国拍摄制作电视剧备案公示的通知," 办公厅, 31 March 2011; available from http://www.sarft.gov.cn/articles/2011/03/31/20110331140820680073.html.

writers to be inspired to historically revise China's history and create an alternate future without the CCP. But there remains a larger impediment to the evolution of Chinese space science fiction, one which may reflect obstacles also found in the aerospace industry, namely a lack of creative thinking.

As discussed earlier, Confucian teaching methodologies have led the current Chinese educational system based on rote memorization. So long as the Chinese government stifles the Chinese people's ability to creatively and innovatively think and write, technological advancements and independent innovation within the aerospace industry will be impeded. The lack of creative thinking is reflected in current Chinese space science fiction.

Chinese space science fiction typically contains four components: aliens, strange worlds, incredible events, and predictions of the future. But, as Liu Cixin points out, space science fiction writing lacks spirituality and wonderment about the mysteries of the universe. Space science fiction typically will use the universe as a backdrop for typical Earthly plots. Stories often encompass self cultivation, personal development, and relationships between people—very Confucian concepts. Until space science fiction in China can create inspiration and awe for the universe, Chinese space science fiction protagonists will be stuck in the roles of fleas and scabies on Pan Gu's skin. This sentiment has been observed in China's human space program. When Col Zhai returned from his famous space walk during the 2008 Shenzhou VII mission, he was asked by a CNN interviewer if he felt different or changed after being in space. Col Zhai emphatically stated "No!" When he returned and saw the video of himself waving the Chinese flag during his EVA, he said he felt "greatness for science" and pride for the Motherland.[16] What a loss for China. Where are the Chinese dreamers? Do they have the capacity to overcome political and social constraints to explore creativity in the aerospace world?

Nonetheless, interest in Chinese space science fiction continues to grow. Part of the reason for the expansion of Chinese space science fiction popularity is due to the establishment of fan clubs that have helped diffuse interest among the people. In recent years, space-related civic groups have been established by Chinese enthusiasts. These clubs allow for like-minded people to gather together and share thoughts and ideas on space science fiction, which, in turn, can help spawn creative and innovative Chinese writings on space-themed stories. The civic groups have risen due to a specific societal desire. However, the central government continues to manage internal social affairs, and the civic groups oftentimes are accused of acting too independent.

While the central government tries to seek "harmony while keeping differences," the government is finding it challenging to manage the cultural impacts of these groups. In March 2011, Premier Wen pushed for an effort to strengthen "social management functions" by the central government and "organize the

---

[16] "Chinese Taikonauts' Great Leap into Space," video from Talk Asia CNN; available from http://edition.cnn.com/2009/WORLD/asiapcf/09/25/china.taikonauts/index.html.

masses to participate in social management in accordance with the law."[17] The difficulties with managing the cultural impacts was evident when the central government attempted to ban alternative futures and time travel science fiction from Chinese society.

Many civic groups abide by the central government's requirement to register for establishment of a new group. For example, the Beijing Unidentified Flying Objects Research Organization (BURO 北京 UFO 研究会) was established in 1994 under the supervision of CAST. With 130 members from government agencies and academic institutions, such as CAST, China North Industries Corporation, and Beihang, the central government has ensured BURO strictly adheres to civil society management. BURO must obtain CAST approval for leadership selection, publications, and conference attendance. If deemed necessary, CAST retains ultimate authority to dissolve the civic group. CAST monitors BURO by having members submit their business and civic management activities to CAST on a regular basis. At the cost of RMB 30 per year, members can gather, discuss, and publish information on UFO sightings, UFO research, and even communicate with unincorporated UFO groups.[18]

UFO popularity has grown to the extent that Liu Cixin suggested adding an alien affairs policy to the CCPCC 12th Five Year Plan (2011–2015).[19] He argued that, as China plays an increasingly important role in aerospace endeavors in the global space community, so, too, should China take the lead in collaboration of alien affairs research and potential contact. He recommended China use its 500 m aperture spherical telescope (FAST) to look for extraterrestrial life similar to the tertiary use of the Arecibo Observatory in Puerto Rico, which collected data that was used by the University of California, Berkeley, to conduct the very popular SETI@Home program. Under CAS, the FAST telescope is being built in Guizhou Province and is scheduled to be completed in 2014. The primary purpose of the FAST telescope is to search for and detect new galaxies and pulsars.[20]

Unincorporated and unapproved space-related civic groups are growing in popularity and are resonating with the Chinese people. Chinese people are joining non-registered space-related civic groups as an outlet against excessive governmental oversight of space-related hobbies that may be perceived as threats against central authorities. While, on the one hand, many provincial governments are concerned with illegal civic groups being established, on the other hand, the "invisible social organizations" (隐形社会组织) are concerned that, once

[17] 顾骏, "提升社会管理的关键," 东方早报, 8 March 2011; available from http://www.dfdaily.com/html/63/2011/3/8/577251.shtml.

[18] To read more on the goals and bylaws of BURO, visit http://www.hudong.com/wiki/%E5%8C%97%E4%BA%ACufo%E7%A0%94%E7%A9%B6%E4%BC%9A.

[19] 姬少亭, "科幻作家建议十二五开展外星生命探测研究," 人民网, 14 March 2011; available from http://roll.sohu.com/20110314/n304293194.shtml.

[20] "世界上最大的单口径射电望远镜: 500 米口径球面射电天文望远镜 (FAST)," 中国科学院, 26 December 2008; available from http://www.lssf.cas.cn/kpzt/gnzz/201002/t20100201_2739087.html.

registered, rights to self-manage the civic groups will be lost.[21] Plus, the Chinese bloggers are quite surreptitious about the formation and activities of illegal, or unregistered, space-related civic groups.

In addition to meeting at registered space-related civic groups, since 1991, space science fiction writers and aerospace professionals have had direct associations, which help strengthen the cultural and creative aspects of the industry. Beihang's former science fiction fan club president Rao Jun (饶俊) became a mission controller for the Shenzhou human space program after graduating from the university.[22] Aerospace scientists and engineers have been invited to participate in space science fiction conferences. For example, on 27 August 2011, Beijing Normal University hosted a conference on the search for extraterrestrial life. The discussion topics included looking for another Earth-like planet and possible visits from aliens to Earth. Again, the FAST telescope was discussed as a means to search for alien communication. The conference was touted as a success to increase interest in astronomy in China.[23]

Aerospace professionals enjoy space science fiction as much as the Chinese people. They have been slowly successful in establishing a bridge between them and the Chinese people via this common leisure activity and interest. The aerospace industry continues to link with fans by using space science fiction as a powerful outreach tool, which, in time, will allow the Chinese people to further identify with future space endeavors.[24]

References

1. Solomone, Stacey. 2006. The culture of China's space program: A peking opera in space. *Journal of Futures Studies* 11(1): 43–58.
2. Guo, Huadong, and Wu, Ji., eds. 2010. *Space science and technology in China: A roadmap to 2050*. Beijing: Beijing Science Press.
3. Tang, Yuankai. 2009. The big dipper guide: China has high expectations for its domestically produced global positioning system. *Beijing Review* 27(9 July 2009); available from http://www.bjreview.com.cn/special/2009-07/03/content_303891.htm.
4. Wei, Long. 2000. China expands space breeding program using recoverable satellites. 29 May 2000; available from http://www.spacedaily.com/news/china-00zb.html.
5. Jim, Dator. 2010. Humans and space: Stories, images, music and dance. In *The farthest shore: A 21st century guide to space*, eds. Joseph Pelton, and Angelia Bukley, vol. 40, Burlington: Apogee Books.
6. Jim, Dator. 2010. Humans and space: Stories, images, music and dance. In *The farthest shore: A 21st century guide to space*, eds. Joseph Pelton, and Angelia Bukley, vol. 40, Burlington: Apogee Books, 39.
7. Dator, James. 2012. *Social foundations of human space exploration*. New York: Springer.

---

[21] http://news.xinhuanet.com/politics/2011-08/11/c_121843159_3.htm.

[22] See foot note 10.

[23] http://www.cast.org.cn/n35081/n35533/n38560/13291843.html.

[24] For further information on science fiction writing in China and other cultures around the world, read the Springer Briefs in Space Development book by Dr. James Dator [7].

Chapter 8
Future Scenarios for China in Space

Nobody can predict the future because it has not yet been written. However, it is always intriguing to try. A productive approach to understanding potential futures is to try to identify and understand the many different images people have of the future, and then identify how those images can affect people's actions, or inactions.[1]

Using what has been covered thus far in this book, four distinct future scenarios for China's space endeavors are offered below.[2] The four scenarios are only sketches of potential futures of what the Chinese aerospace industry may look like in a continued growth scenario, a disciplined society scenario, a collapsed society scenario, and a transformational scenario. Beyond these four scenarios, the future that emerges can still manifest in innumerable forms.

The purpose of this chapter is to introduce novel ways to think about China's future in space, to foment discussion and communication between the Chinese aerospace professionals and Western counterparts, and to increase understanding of China in space. These futures raise two key questions: Where is China's aerospace industry heading, and, will Western space programs be a significant part of China's journey into space?

Future 1: A Future of Continued Growth

A future image depicting China's aerospace industry on a continuous path of growth is very common. In fact, it is a tool frequently used by Western analysts. Using trend analysis, China's already well-developed growth in the industry is traced back to its historical evolution, and then future actions and advancements are extrapolated based on the rate of development assessed in the past. This continued growth scenario represents government-sanctioned extrapolation

---

[1] For more on futures studies, see Jim Dator, "Methods in Futures Studies," (lecture notes presented at University of Hawai'i at Manoa, 1985) and see [1].
[2] Based on Jim Dator's work on the four archetypes of futures see [2].

S. Solomone, *China's Strategy in Space*, SpringerBriefs in Space Development, DOI: 10.1007/978-1-4614-6690-1_8, © Stacey Solomone 2013

of China's future in space, which currently is creating a foundational base with a very long-term vision of where China is heading in terms of space technologies.

In this future scenario, the Chinese central government has successfully and fully integrated the civil, military, commercial, and academic sectors of the space programs, and has merged them into one seamless aerospace enterprise. Despite foreign government complaints regarding military involvement in the Chinese space station and lunar exploration programs, and only slight domestic resistance from CASC and CAS, the large-scale mandated reorganization of the institutional orgware has been accomplished in only a couple of years. The new organization, named the China Aerospace Enterprise (CAE), is able to satisfy the civil, military, commercial, and academic requirements by the central government for satellite communications, satellite timing and navigation, and imagery demands.

Crucial to this merger is the critical need for imagery from Earth observation satellites to map foreign resources of untapped oil around the globe to feed China's ever-increasing energy needs. To retain internal stability by providing new sources of energy, the CCP and PLA push out beyond China's borders. CAE is required to provide reliable and consistent advanced indigenous satellite navigation in the face of regional unrest over disputed oil-rich areas and to remove dependence on the United States for global navigation satellite services. Chinese demands also grow for more satellite communications bandwidth to support military operations reaching outside its littorals as they ensure Chinese access to newly exploited energy sources. More satellite communications bandwidth also supports domestic economic opportunities. With full integration of all sectors of the aerospace industry, China is able to maximize use of its on-orbit constellations within a short period of time by removing redundancies in once-stovepiped space programs.

Because of the great success with the establishment of CAE, the aerospace industry is able to continue supporting the central government's fast "China Speed" approach toward space programs as first promulgated by Deng Xiaoping. In addition to short launch cycles allowing rapid access to space for more satellites on-orbit, the increase in on-orbit satellite constellations leads to greater economic opportunities. The commercial arm of CAE is allowed to pursue profit making from the Beidou navigation satellites, communication satellites, and imagery satellites, thus making the Chinese aerospace industry an economically viable source of income. This leads the new organization to become less dependent on central government monies and to look less like an SOE and more like a Japanese keiretsu with Chinese characteristics.

Because it is now less reliant on government-generated monies, CAE exerts a more rigorous role in the direction of future missions in space. The central government recognizes its decreased ability to wield domestic "soft power" for controlling the prolific space programs. Instead of cracking down on CAE, the central government decides to embrace the change and welcome the keiretsu-managed organization, just as it weathered the relationship with ZTE in the telecommunication industry. The CCP encourages the formation of a symbiotic relationship where CAE is granted limited decision-making authority and, in turn, is able to introduce innovative managerial leadership options to the CCP.

As Chinese aerospace professionals make great leaps in space technologies, it becomes apparent that the weakened power of promulgation to redirect their work is, perhaps, more in line with political goals but not with scientific or technical objectives. Such processes are less effective and result in blocking further technological innovation. The Chinese keiretsu-style of CAE allows effective negotiation of internal space-related issues with the CCP, while also presenting a single face to the global space community. Some challenges remain, however, such as working around the old patron-client system, prolific corruption, and favoritism within CAE. For example, CASC, albeit infrequently, encounters limited insubordination from non-Beijing units such as SAST and XSCC. But, these minor infractions are overlooked with the right favor-currying tactics.

In this future scenario, China's aerospace industry abounds with opportunities for the aerospace professionals who hold the most lucrative government jobs with great benefits and job security. The scientists, engineers, and technicians are respected, have rapidly reached upper middle class status, and are able to accept the occasional favors from their network of close associates. Their clout in the global space community supports their beliefs that they are morally just and superior over their Western counterparts. Internally, their patriotism lends to tolerance of the friendly socialist system under which they work, knowing that the CCP is communist in name only. Passing the leadership roles to younger age cohorts remains a smooth process. Just as the 4th generation of aerospace leaders nurtured 8th generation professionals into leadership roles, so, too, do the 8th generation professionals support their younger compatriots of the even more technologically savvy 9th generation. Touted as the new Mandarins of the global space community, the 9th generation of very smart, very capable young engineers will lead the world's space activities with China at the helm for decades to come.

Domestically, spin-off technologies diffused in society and pride by the people in the space programs continue to provide limited cohesiveness in domestic matters. The same pride helps to extend continued support for the PLA and CCP. Unintentional spin-offs are relatively harmless to the flawless image of the monolithic CAE and are chalked up to being frivolities no more than "spaceship video games" and space-themed elixirs.

Internationally, nations with developed space programs begin to fear the rapid rise of China in space, yet can only conform in order to work on the many Chinese space programs. These include a Chinese space station, exploratory probes for space sciences, and satellite programs for practical applications that provide foreign nations with imagery, navigation, and communications services (at a price). Because aerospace leadership decision-making processes are founded in China's ancient philosophies of Confucianism, Buddhism, Daoism, and Legalism, their dichotomous nature and dialectic thinking remain very effective and feed the notion that they indeed should be at the helm of the global space community, as they are moralistically correct. Under this scenario, CAE paves the way to space for decades to come as a monopolistic space mega-power on the global scene.

This future scenario is largely a government-sanctioned view and, in this author's opinion, also the least likely scenario for China's future in space.

Most government-sanctioned plans and policies rely upon a continued growth future based on extrapolative methodologies. This future scenario does not factor in the emerging and unaccounted for events or technologies that can and will alter China's space future.

Future 2: A Future of Disciplined Control

This alternative future foresees that a "disciplined society" is created. This occurs when a regime is focused on sustainability, such as economic or environmental sustainability, and forces a slowdown to stave off the perceived limits to unsustainable growth. The political power in charge may use draconian measures based on ideology or military might to enforce this slowdown.

In this scenario, the Chinese central government, determined to retain legitimacy and secure internal stability, steps up to the challenges it has superficially faced for decades. These key factors include: an overheated economy; massive environmental pollution; land and resource scarcity; growing elderly population and imbalanced birthrate of boys and girls; countrywide ethnic minority riots; governmental corruption; and uneven growth in social classes and provincial powers. These largely negative forces compel the central government to use draconian measures to ensure internal stability. The space programs do not fare well as the central government falls back on similar actions taken during the Cultural Revolution, namely, a reorganization of the aerospace industry that placed space resources directly under military control. This time, the PLA is given complete authority over the aerospace industry, including decision-making power for the missions of future space programs.

At the same time, the central government reduces the diffusion of space technologies into society to a manageable and easily monitored pace by re-invigorating the power of promulgation and strengthening the propaganda machine to instill a more conservative ideology for the Chinese people, aerospace professionals, and the military. Ideology and political correctness are the main tools used by the propaganda department of the central government. Internally, the aerospace industry is not powerful enough to provide the economic responsibility needed to counterbalance reduced monies from the central government, nor can the space programs create and maintain economically viable businesses. However, the Chinese people, the aerospace professionals, and the military are willing to be organized around sustainability as a new focus for the good of the motherland.

Culturally, the Chinese people want to avoid being thrown into an economic crisis and looming chaos. The linchpin to the central government's success is the PLA. The Chinese people and the aerospace professionals are willing to rally around the People's Army. In fact, they do just that. With full integration of civil, military, commercial, and academic sectors underway for years, only a relatively minor reorganization is required to take place to put the PLA solidly in control of all space programs. During the Cultural Revolution, where factories were moved to the Third Front—resulting in a rather unhappy CAS, which lost much authority in the space sciences field—this reorganization encounters little resistance as it becomes evident, via the propaganda machine, that the path of rapid growth in the space programs is not sustainable. This transition is smoother, quicker, and justified. PLA leaders are put into positions of authority within the aerospace industry, including universities, such as Harbin Institute of Technology, Tsinghua University, and Beihang University, which are key to developing extensive space technology R&D programs.

Growing decentralization and imbalances of power between the provinces threaten internal stability. The CCP, unable to solve provincial loyalties that exist within the PLA, attempts a military reorganization, which alleviates concerns over military loyalties. Within the aerospace industry, military leadership shuffling results in consequences, as the officers lean heavily on consensus building and pre-existing military guanxi for decision-making on space-related issues. With heavy reliance on guanxi comes more favor currying and corruption. The PLA manages to get projects done, albeit at a slow and methodical consensus-building pace, and the direction of the space programs changes. With the PLA squarely in charge, space programs are geared toward supporting China's expanding military might. Military leadership seeks to allocate energy resources from disputed territories.

The PLA also seeks new energy resources from nations willing to partner with China, such as those in Africa, to serve increasing domestic energy demands. In order for the PLA to access both the disputed and untapped foreign resources, the military redirects most of the aerospace industry missions toward satellite communications, navigation, and imagery to support these endeavors. Overhead imagery helps map out potential resources around the world. Remote sensing also serves the unsavory task of supporting domestic law enforcement needs. Satellite navigation and communications help the PLA reach BLOS to acquire these new resources, and, subsequently, support the PLA Navy's desire to become a legitimate deep blue water navy. As the PLA acts as glue for the CCP to retain internal stability, all technological advancements within the aerospace industry are geared to support the PLA.

Once the space programs are redirected to support the navigation, communications, and imagery needs of the military, they reach a point of growth that suits military demands and thus begin to stagnate in terms of innovation. The CCP recognizes and supports this occurrence, since it legitimized scaling back the space programs that never did generate much profit for the PLA. Some diffusion of space technologies continues, such as the commercialization of Beidou, but large non-military programs suffer, such as the space-based X-ray telescope. Programs focused on scientific exploration of the universe are cut. Deep space exploration and space sciences are once again greatly curtailed, as was the case during the Cultural Revolution, to CAS's demise.

The space programs are now looking inward to support the motherland and the military, and are not geared toward sinification of the global space community. In fact, the only international reach is to gain monies and technologies from foreign space programs, as was experienced during the 1980–1990s. The CCP and PLA are grossly aware of implicit values from foreign space programs infiltrating China's domestic space programs, and so they proactively limit this source of influence. New ethics are promulgated to affect the aerospace professionals—ones which tout nationalism, political correctness, and moral superiority over Western values. Space ethics are pushed via promulgation and mandatory training sessions as the entire country concentrates on solving domestic problems.

Nobody wants a collapse of the economy, and aerospace professionals know this is a necessary reform taking place across all industries in China. They know it is their duty to "eat the bitterness." Even when aerospace professionals occasionally offer better innovative management methods, the PLA clamps down on such ideas, labels them foreign, and stifles the engineers and scientists, tempering their motivation and redirecting their goals toward serving the motherland. The engineers and scientists learn it is better to promote a practical and slow-paced aerospace industry than to stick their necks out against the military stalwarts in power. The PLA finds 6th generation aerospace professionals especially willing to follow the military's lead, and subsequently places them in management positions.

The Chinese people throw complete support behind the military-controlled aerospace industry thanks to the very successful propaganda machine. National space heroes are also heroic military men, and foreign space endeavors are looked upon with suspicion. Technological advancements diffused into society are tempered at a pace the PLA can monitor and control and which satisfy the people's desires for better standards of living and life amenities. This future remains in place for the next few decades. In this scenario, the aerospace industry is one of many tools effectively used by the CCP and PLA to solve their rather serious internal threats to stability. But, what if the CCP and PLA are not able to handle those internal threats? What becomes of the aerospace industry in that scenario?

Future 3: A Future of Collapse

The decline of an empire usually comes quickly and is often the result of a variety of ill-timed events. Such is the case for China under this scenario. After a long and successful rise to power, quite suddenly China falls. With mounting economic, social, environmental, and political problems looming large over the CCP, in a last ditch effort to retain internal stability, a new set of radical policies is introduced. Economically, the CCP knows it has reached the cusp of sustainable growth and implements a number of policies that hurt the lower class, retain the middle class, and allow the upper class to explode with affluence.

The upper class is able to use financial resources and guanxi to prop up the banking system. However, once they realize that the house of cards they had helped create is teetering, they move their monies en mass from the Chinese banks to overseas accounts. When the banks fall, the middle class is hurt and the peasants take to the streets in rolling riots across the country. Social instability results, calls for leadership change ensue, and widespread social disorder leads to a breakdown in cross-provincial transportation caused by homemade pipe bombs on railroad tracks. With local guanxi supporting local cities and towns, the breakdown of inter-provincial transportation of goods leads to food shortages in the poorer provinces. PLA assistance is stretched too thin, as the crisis spreads across several inner provinces. Qinghai is especially hard hit. There, satellite-derived communications are blocked as rumors of village-level starvation spread. The localized famines foment further provincial fragmentation. The other provinces are able to thwart food shortages, but are not able to escape local revolts and a slow decline into poverty among the rural population.

With the environment overloaded, people drink polluted water, diseases spread, and the overall health of the people decline. Stressed hospitals treat only the people who can pay the highest fees for care—a situation that snowballs into further riots. Coal mines are not mined as the workers protest, and foreign resources are not sufficient to meet the national consumption of energy. Intermittent blackouts occur across most of the major and secondary cities across China, as satellite imagery is able to capture the internal chaos in nighttime images, depicting clear delineations between provinces with power and those without. Foreign satellites also image the millions of Chinese diaspora, who consist mainly of hungry villagers spilling into neighboring countries such as Cambodia, Vietnam, Russia, and even North Korea for food, shelter, and escape from chaos, raids, revolts, and random murders and lawlessness. Foreigners either willingly leave or alternately are expelled as the CCP tries to find a foreign source to blame for the collapse.

The PLA mercilessly squashes uprisings and attempts to establish order in what appears to be a society descending into chaos. As was the case during the 1989 Tiananmen Massacre, military units from other provinces are called in to do the dirty work. This results in factions between the military units blaming one another for killing people in their home provinces. The military is partially successful in putting down the revolts, but as local military leaders realize they can help their local provinces and, at the same time, make a profit for their assistance, military leaders also succumb to the temptation of corruption and self-preservation.

Space-related technologies that once filtered through most sectors of society, such as satellite communications and broadcasting, are now used for localized political gain and propaganda or are shut down because of central government censorship control. Most of the missions of satellite programs deemed necessary for military missions are replaced with other technologies that serve the same purpose, such as domestic fiber optic communications and use of near-space dirigibles, or high altitude UAVs for imagery over contesting provinces.

With a rising dependence on local patron-client systems, the professionals in the aerospace industry quickly learn that if they are to survive, they need to attach themselves to the local military leaders. Therefore, many of the space programs that once crossed provincial boundaries are stymied as lengthy negotiations take place. While the engineers and scientists strive to survive what is now becoming a second decent into chaos since the birth of the space industry in China, lower class Chinese look for a source to blame for their decline into poverty. The middle class professionals make an easy target, and, the CCP is glad to have the intellectual class as a scapegoat for the people's discontent. The people even plunder Xi'an Satellite Control Center, which results in the shutdown of a critical node for satellite control and monitoring of on-orbit constellations. Also, a major railroad transfer point between Beijing and Jiuquan SLC is bombed and blamed on Uyghur separatists, even further slowing down a stagnated launch capability. Aerospace professionals work in whatever capacity they can to put food on the tables for their families. Some try to flee the country, resulting in a brain drain; a few noble others try to press for the survival of the industry, but to no avail. Although the aerospace industry was able to survive the Cultural Revolution, in this collapse scenario, it does not survive.

Future 4: A Future of Transformation

In a transformational society, there is an end to current forms and emergence of new forms of beliefs, behaviors, and organizations. The transformational society offers a complete paradigm shift based on future technologies that leads to new values and behaviors. Because it is so fundamentally different from what we witness today, Dator's Second Law of Futures applies: Any useful statement about the future should seem ridiculous. To illustrate, imagine trying to explain the Internet and all of its primary and secondary effects on society to a villager in China in 1949. It is precisely because technology is a major agent of social change that such a fundamental technological shift—a tsunami of technological change—transforms society (values, behaviors, organizations).

In this scenario, a technological transformation occurs within China's aerospace industry that forever changes the face of the world. China is able to make space economically viable with the creation of a single space technology. China desperately needs new sources of energy to feed domestic demands not only for an increasing population but also for hundreds of millions lower class emerging into a middle class who expect higher standards of living.

At first, the Chinese aerospace industry uses its satellite imagery constellations to map out potential foreign energy resources in disputed territories in the Asia Pacific region and untapped resources in Africa. Satellite imagery helps discover yet untapped oil in the Spratly Islands, but that comes with a high military price that gives only limited payoff. Other sources are exploited with similar results. While detection and exploitation of Africa's resources feed China's energy demands for a couple of decades, it does not last. Internal strife in the Horn of Africa limits China's access to these resources and provides only a small percentage of the amount of energy China needs. The military is spread thin around the globe trying to ensure China's access to energy resources.

China looks to alternative energies and hopes for successes similar to those experienced with hydropower damming projects. The world's largest wind farms are established in Xinjiang. China establishes solar panel farms spanning the entire Taklamakan Desert in the same region. Nevertheless, it is a small group of forward

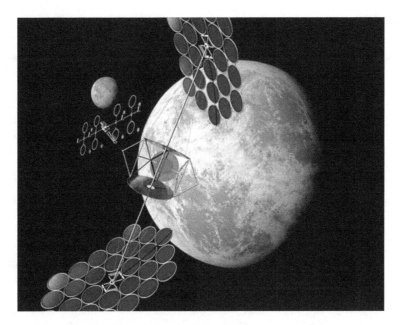

Fig. 8.1 Artist's depiction of an SBSP collector. (Image courtesy of NASA.)

thinking aerospace scientists in CAS who offer the CCP a plan to explore space-based solar power (SBSP) options. Space-to-ground solar energy has been a theory bounced around the international scientific community for decades, but the projects were previously considered to be too expensive and too risky to take to fruition. The CCP, however, embraces the plan and allows CAS to assume control over several facilities under CASC to create the SBSP program.[3] The military is encouraged to offer space-related facilities under its control as it sees the potential for military applicability once the technology is brought online[4] Fig. 8.1.

With such dire needs for energy, the Chinese central government redirects massive financial resources to support the megaproject under the 863 Plan. Within a reasonable amount of time, under CAS leadership, the SBSP program is launched into orbit and the components to the solar collector are assembled in situ. The CCP decides to use most of the region in the Taklamakan Desert as the receptor site for the microwave transmitted energy from the on-orbit collector. The CCP also agrees to foreign involvement and allows foreign space-faring nations to directly receive energy from the Chinese collector in orbit at a very high profit for China. A number of foreign nations then build ground-based receptors within their borders (although many countries have trouble finding enough isolated land for such a dangerous site).

---

[3] References [3, 4].

[4] Image from http://www.power-technology.com/features/featuregiant-leap-for-space-based-solar/featuregiant-leap-for-space-based-solar-1.html.

For non space-faring nations, China agrees to transfer energy via a ground-based grid, again, at a very high profit for China. China makes space profitable beyond expectations and dreams. For the first time, space truly becomes industrialized, and the world's energy problems are alleviated, a result that catapults China to the status of prime superpower of the world. With this transformational technology come changes to China's political and economic status in the world. This technological revolution changes Chinese politics, changes CAS's status in the aerospace industry, and changes the standard of living for the Chinese people. The world broadly recognizes China for its innovative and transforming space technology.

With new technology transfers comes implicit knowledge and influence. Under this scenario, China is able to use the SBSP program to culturally sinify the world with Chinese ethics and morality. The CCP's sinification plan lifts its status in the world and once again increases its influence with other nations, as it did under the old tributary system. The CCP, knowing its values are morally superior, seeks to better the world by influencing cultural development in line with Chinese ethics, values, and behaviors. The success of the SBSP program is used to support this vision of China's once lost status in the world based on a two-pronged approach. One is to provide free international satellite broadcasting equipment to the world. The second approach is to combine the equipment with free transfer of data via a series of Chinese routers in space.

The cultural sinification of global educational systems proves to be swift. With free broadcasting of lessons in the Chinese language geared toward school-aged children in K-12 grades, schools around the world are happy to integrate this new form of technology into their curriculum. With language comes culture. The successful Confucian Institutes reduce or eliminate their tuition fees as a part of an agreement with the CCP. Visiting English-speaking professors are invited to experience China's new English village to help set up similar Chinese villages in America, Europe, and Australia. Confucian morals began to creep into Western societies. Chinese methods of raising children become more popular. Western high school and college students begin to debate governance theories, such as rule of law versus rule of noble leader, and to challenge traditional Western values based on individual freedoms.

The very cleverly created lessons broadcast around the world into children's classrooms bring with them Chinese values and ethics, and garner much international support for the CCP in terms of international prestige and global economic investments. The process is gradual and goes unnoticed as a threat to traditional Western ways of thinking. The lure of good quality and free education wins out, and China is able to exert vast influence over the entire world. A complete technological transformation has taken place creating a world one would not be able to recognize today. But, nothing lasts forever....

The purpose of these four distinct scenarios is not to predict the future but rather to foment thought "outside the box" and allow for a wide range of speculation. These widely disparate scenarios allow one to consider the future role of China's aerospace industry within the larger context of China's strategic goals, potential threats, and vulnerabilities. It is important to understand why China is going into space and if Westerners will be a part of its journey.

References

1. Dator, Jim. 1998. The future lies behind! Thirty years of teaching futures studies. Introduction to the special issue on teaching futures studies at the university level. *American Behavioral Scientist*. From http://www.futures.hawaii.edu/puplications.html.
2. Dator, Jim. 2009. Alternative futures at the manoa school. *Journal of Futures Studies* 14(2): 1–18.
3. Ji, Bian. 2011. Technology roadmap for space solar power proposed by Chinese scientists. *Aerospace China* 12–13.
4. Gao, Ji, Hou, Xinbin, and Wang, Li. 2010. Solar power satellites research in China. *Online Journal of Space Communication*. From http://spacejournal.ohio.edu/issue16/ji.html.

Chapter 9
Top 10 Things to Know About China in Space

1. *Current Developments in Capabilities*: China's current developments involve on-orbit, launch, and ground segment capabilities. First, it is important to understand basic on-orbit capabilities in the following categories—satellite communications, satellite timing and navigation, ISR, environmental monitoring and disaster management, scientific and human spaceflight efforts, and offensive and defensive counter space. Second, China's fast-paced launch capabilities support its aggressive launch plans and access to space. China's fast launch cycles are able to thwart delayed satellite launch scheduling. Also, better SLC locations relative to orbit insertions mean lower costs for launch services. Third, Chinese ground control capabilities at XSCC, BACC, and the fixed and mobile (land and sea) TT&C sites are crucial to maintaining an active space presence. In addition, satellite tasking, collection, processing, exploitation, and dissemination of data take time, organization, and leadership management to determine prioritization of customers. As the ground control capabilities improve, so, too, will China's space situational awareness (SSA) detection and tracking capabilities.

Exploring current technological developments allows insight into why, how, and when China will develop and use its hardware capabilities. For example, Chinese satellite bus modularization R&D will pave the way for potential on-orbit servicing (OOS) to extend satellite lifetimes.[1] The building up of the Beidou constellation will result in a fully operational global navigation system and may result in changes to the military's operational plans as the PLA moves away from perceived dependencies on GPS. China's current development of particular Earth observation satellites may help identify world geological sites they may be more interested in for, perhaps, exploitation of natural resources.

The Chinese aerospace industry's R&D into hardware capabilities supports national strategic goals. It is critical to understand these R&D efforts to develop

---

[1] OOS would use a type of plug-n-play satellite module that could be removed and replaced on orbit, say, for extra battery life or an extended payload capability.

S. Solomone, *China's Strategy in Space*, SpringerBriefs in Space Development, DOI: 10.1007/978-1-4614-6690-1_9, © Stacey Solomone 2013

new systems capabilities to gain an understanding of China's current strategic focus in terms of military, economic, and political directions. For example, if China should build large, nuclear-hardened satellite constellations reminiscent of the Cold War era or small, disposable, and redundant satellites reflecting a more modern trend, then Westerners can better understand Chinese concerns and risks and, thus, be able to better communicate with their Chinese counterparts on these strategic issues. An accurate assessment of how, why, and when cannot be done without first knowing basic hardware capabilities.

2. *China's Space History*: China's unique entrance into the space arena shaped how, why, and when aerospace developments occurred. The Cultural Revolution had a significant impact on the space programs and the pace of scientific and technological developments within the aerospace industry. On the one hand, foreign technologies were no longer entering China, nor were scientists and engineers cultivated to support the growth of indigenous space technologies. On the other hand, certain sectors of the aerospace industry were financially supported during a time when there were very few economic resources so long as they were deemed vital to national defense and international prestige efforts.

Looking at the history of the meteorological satellite program, China's developments in space-based meteorology influenced its current level of proficiency in the Fengyun program. During the Cultural Revolution, Premier Zhou encouraged China's aerospace engineers to develop indigenous meteorological satellites while still using limited data from foreign satellites. By the end of the 1960s, China developed its first meteorological satellite receiver to obtain images from NASA's Environmental Science Services Administration Satellite. By the end of the 1970s, China's own designs for the Fengyun constellation were completed [1]. At the same time, Mao encouraged the Chinese people to set up weather monitoring stations in communes across China, which resulted in as many as 16,000 weather and rain stations. This, in turn, diffused the technology into society and connected it directly with the people by linking weather monitoring with national pride. If weather had not had such an historical significance and deep cultural association with political power, satellite meteorology likely would not be at the advanced state it is in now (Fig 9.1).

Historical aerospace endeavors have greatly impacted China's current technological capabilities and pursuits, and, in turn, as seen under the leadership of Deng, also have affected the organizational structure. Because the orgware of the aerospace industry underwent massive restructuring first under Mao and later under Deng, it is quite feasible that a similar large-scale reorganization may take place in the near future. Such restructuring changes the aerospace industry's values, missions, and capabilities development. By learning from China's space history, one can learn that the potential exists for another overhaul of the entire industry. Only by understanding Chinese space history can one look for the signs of change in the near future.

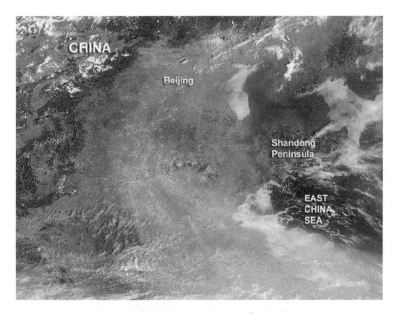

Fig. 1.1 Satellite imagery of weather patterns over China.[2] (Image courtesy of NASA)

3. *China's Space Future*: According to the 2011 Space White Paper, China intends to explore space, Earth, and the cosmos; utilize space for peaceful purposes; and promote human civilization and social progress. The rather lofty goals in space are not matched with what the Chinese may actually achieve. This is due to the realities and limitations of physics, time, and resources.

   To illustrate, in the 2011 Space White Paper, China indicates plans to develop and implement a Chinese-style integrated Earth observation system to connect the technology components (hardware, software, and orgware), which may resemble something akin to the U.S. Operationally Responsive Space (ORS) initiative.[3] To accomplish this goal, China will pursue a number of avenues. China is working on modularization of Chinese satellite buses, which may result in successful OOS capabilities. China plans to develop the CZ-6 high-speed response launch vehicle capable of launching a 1 ton payload into a Sun-synchronous orbit at 700 km altitude. China plans to develop the EOS infrastructure that would integrate data from various Earth observation satellite constellations—Fengyun, Haiyang, and CBERS—at the ground stations to improve data fusion capabilities. China also

---

[2] Image from http://www.nasa.gov/mission_pages/fires/main/world/china-20120612.html.

[3] The ORS concept is to provide rapid augmentation, reconstitution, and exploitation of space capabilities by facilitating rapid assembly, integration, deployment, and operations of space assets into the current space architecture. Through collaboration with the broader space community, ORS also explores new command and control, acquisition, business, and TPED processes for space technologies. For more on ORS, visit http://ors.csd.disa.mil/about-ors/index.html.

plans to integrate reception, processing, archiving, and distribution of satellite data into a centralized system by a multi-functioning TT&C network.[4] Having modularization, rapid launch, Earth observation data fusion, and multi-functioning TT&C capabilities resonates with goals analogous to ORS, at least within the Earth observation mission set. Yes, lofty goals indeed, but can they be achieved? How will personnel at the integrated data ground stations determine prioritization of competing customers' requirements, what and when data is collected, and by which of the Earth observation satellite constellations? In this example alone, China faces many technological, organizational, and institutional difficulties and challenges to overcome in this future space plan.

China's space programs are not determined by a single, monolithic entity. Nor does the Chinese aerospace industry have only a single vision of its future in space. Decisions are made by a slow and methodical consensus building method, which potentially will increase as a new leadership group comes to power in the Politburo. Chinese plans are subject to change with each new Five-Year Plan. Leaders in decision-making positions likely must vie for resources, and negotiate client frictions among aerospace planners and program managers within a very complex patron-client system. Also, Chinese engineers and scientists will continue to run into technological and financial resource limitations. Yet, at the same time, the Chinese aerospace industry sits poised—on the cusp—of accomplishing great feats in space in the near future. This leads to the next thing to know about China in space.

4. *On the Cusp*: China's aerospace industry is on the cusp of independent innovation. There are four components essential to establishing a foundation for truly independent innovation, some of which are already well developed.

First, CASC has eight large space R&D centers and production facilities and 12 national key laboratories. In addition, it is partnering with academia and commercial enterprises to create a solid foundation in aerospace sciences and technologies ripe for the creation of new technologies. CASC realizes it cannot continuously leapfrog technologies because this means it will have to follow foreign strategies and processes in the aerospace industry. Instead, CASC is seeking independent paths to innovative technologies.

Second, the aerospace industry has made significant progress in exporting products and services to the international market, such as sales of satellites, launch services, and satellite application services. Although there have been setbacks, such as NIGCOMSAT-1 for instance, these efforts are growing and succeeding. China is also pursuing standardization methods within the international space communities. Here, China faces an uphill battle with implementing its standards within an already established and mature global space community, but continues to make slow and methodical headway in this field.

---

[4] Guo and Wu [3] and Information Office of the State Council of the PRC [4].

Third, CASC is beginning to explore limited domestic market freedom. By the end of 2009, CASC claims to have made a total profit of over RMB 7 b. They still have a long way to go, but with the central government's push for CASC to find other resources to support the space programs, CASC should be able to take advantage of this and reduce its dependence on central government monies.

Fourth, CASC is embracing the cultural tradition of dialectic thinking and innovative managerial methods for decision-making with Chinese characteristics. Chinese-style negotiations are not new to Western businesses; however, there still is not enough understanding of the differing negotiation styles between China and foreign space-faring nations and/or adversaries, as was illustrated in the Protostsar-1 case. Time is limited to work with China's aerospace industry as it exists today. Once China begins independently innovating space technologies, foreign engagement with China's space programs will need to radically adjust.

5. *Achilles' Heels*: As discussed throughout this book, China's aerospace industry is on the cusp of achieving genuine independent innovation. With this strategic space goal comes a plethora of vulnerabilities found specifically within the aerospace industry. First, as mentioned above, so long as the Chinese central government continues to promote a policy of leapfrogging technologies based on acquisition of foreign aerospace hardware capabilities, China's space programs will remain in a state of catching up to foreign space-faring nations. But, this will not be an enduring problem for China, as Chinese aerospace professionals seem to have the drive to break through these technological barriers in the near future.

Second, as the central government continues to foster a favorable environment for aerospace professionals, it has the negative effect of creating a new social class that contributes to the growing disparities between the emerged middle class and the majority of China's poor population. Economic class struggles could pose a looming threat to internal stability and Party legitimacy. Despite being aware of this threat, the CCP appears to be slow in raising the standards of living for the rural poor at a similar rate as that of the urban middle class. This puts the aerospace professionals in a precarious situation. On the one hand, they readily embrace buying real estate, automobiles, and imported Ikea furniture. On the other hand, as Chinese history dictates, the middle class intellectuals make for very convenient scapegoats if the central government needs to squash internal turmoil, riots, and protests due to domestic economic disparities.

Third, the organization of the aerospace industry is a constraint on development of new technologies. Integration of the military, civil, commercial, and academic sectors will help alleviate resource allocation limitations, foster enterprise clusters and market viabilities, and break down stovepipes.[5] But, much of the endeavors in the civil, military, commercial, and academic sectors remain isolated from one another, which stifles R&D information sharing and leads to redundancies.

---

[5] Information Office of the State Council of the PRC [4].

The central government recognizes this problem. But if these sectors are integrated within the aerospace industry, this could lead to new vulnerabilities as China faces increased openness and transparency. Also, more openness in China's aerospace industry may lead the PLA to once again enjoy the benefits of profit making reminiscent of the 1990s, when PLA, Inc., seemed almost to spiral out of control. China's President Xi, could be greatly challenged to divest the PLA from economic endeavors like Jiang Zemin was able to accomplish in 1999.[6]

Fourth, the Chinese aerospace industry may be taking on too many space programs. In addition to the space programs already underway, in the next 5 years, they also plan to:

- Develop three new space launch vehicles (CZ-5 heavy lift, CZ-6 high speed response rocket, and the CZ-7 that will be able to place 5.5 ton into SSO);
- Improve Earth observation constellations including synthetic aperture radar and high resolution Earth observation satellites;
- Launch new communications satellites, recoverable satellites, and a space-based X-ray telescope;
- Implement expanded regional positioning, navigation, and timing services from the Beidou constellation;
- Launch the Shenzhou X to continue to conduct manned rendezvous and docking with the Tiangong-1 spacelab;
- Develop plans for a manned lunar landing and extended taikonaut stays in LEO onboard a Chinese space station;
- Strengthen SSA work on orbital debris and spacecraft protection; and
- Pursue an Earth observation "system of systems" as discussed above.

In an attempt to prioritize missions, the 2011 Space White Paper specified that space applications take priority over space science and exploration.[7] Yet, the central government is directing the aerospace industry to simultaneously develop a lot of space programs in the space sciences and exploration area. While the central government is guaranteeing steady financial investment for space activities via a new multi-channel space funding system, it also has ordered the aerospace industry to explore ways to make space profitable by pushing for satellite communications, navigation, and ISR constellations to become industrialized, commercialized, and economically viable [4].

Even if the industry did foster innovative measures to make satellite services profitable and counter the extremely high prices to develop such systems, to what extent could the aerospace industry balance central government and military customer requirements with those of potential commercial customers? For

---

[6] *Note* Although Jiang was successful in divesting the PLA from business ventures, it is interesting to note that the telecommunications industry continued to have military guanxi. And, yet, whereas military involvement in business ventures was blamed for decreases in readiness and increases in corruption, the telecommunication industry is at the forefront of China's successes in independent innovation.

[7] Information Office of the State Council of the PRC [4].

example, how will the aerospace industry convince banks to use Beidou for timing services for ATMs and ensure reliability when GPS is already a proven, reliable, and free technology? With mounting resource demands by the plethora of space programs, how will aerospace leadership be able to balance national security needs, develop new highly technical space programs, and open space to commercial use? It does not seem likely China will be able to accomplish all of these goals.

6. *Chinese Perceptions of their Achilles' Heels*: In addition to the vulnerabilities discussed above, based on recent activities in the aerospace industry, there are indications that the Chinese central government is concerned with additional perceived weaknesses. First, the aerospace industry has been directed to strengthen work on technical aspects of the space programs, such as the independent navigation and timing system, indicating leadership is concerned with dependencies on foreign satellite services. Second, China wants to be a bigger player in the global space community and is promoting Chinese standardization practices so that it can use this as leverage in the near future. Third, China seeks to improve the domestic aerospace legal structure and be able to vie with the global space community's established legal norms. Fourth, China has stressed that the space programs must support the long-term goal of informatization (信息化) of the PLA, which puts pressure on the industry to support the military and, at the same time, integrate with commercial and academic sectors to become economically viable. Sixth, not only are the sectors of the aerospace industry having difficulties integrating, but the overall orgware is not able to keep up with advances in space technological hardware and software. Lastly, the potential for corruption is increasing as more actors are being folded into the mix and disparities between old stalwarts and new visionaries for China's place in space are rising to the top. A better understanding of how China views its own weaknesses and perceived vulnerabilities can lead to better understanding of why China is pursuing particular goals in space.

7. *Aerospace Leadership Decision-Making*: Decision-making processes are influenced by space ethics, which permeate the aerospace industry. Chinese space ethics manifest in China's dialectic thinking and dichotomous approaches to decision-making. Dialectic thinking and dichotomous approaches embrace many complex elements that, while different from one another, are mutually dependent for their existence and are in continuous evolution. The complexities found in the concepts of harmonious flux and cultivation of benevolence can be found within these approaches to decision-making.

Decision-making by the Chinese aerospace leadership is a slow, methodical, and pragmatic process. Leaders carefully analyze problems and work very slowly to conceptualize challenges and build consensus. Conceptualizing the challenge is a part of the dialectic thinking process. Consensus building takes time and is rooted in Confucian ethics, which stress the group's good over that of the individual and support the moralistic notion of inclusiveness and acceptance. Most importantly, decision-making is an iterative process. However, what at first appears to be a detriment to the need for speedy decisions in a time of crisis, has, in fact, worked well for the Chinese aerospace leaders.

Today, we see three forms of decision-making processes within the aerospace community (rule by reputation, rule by consensus building, and rule by law). All three leadership styles have roots in Chinese ethics. Rule by reputation is based on the Confucian model of an ethical nobility ruling by example and demonstrating how one ought to conduct oneself when in power via moral exhortation. This has manifested in recent history in the form of charismatic leadership from Mao and Deng. Rule by consensus building has roots in Daoism, in which leaders should perform, or not perform, actions to maintain a balanced status quo and retain harmonious flux. This can lead to inaction in cases where action is needed. Lastly, rule by law has roots in Legalism, which advances the idea that institutional structure is needed to influence people to behave a certain way via legal coercion.

8. *Age Cohorts of the Aerospace Professionals*: Within China's aerospace industry exist several age cohorts, or generations, of professionals—scientists, engineers, and technicians—who share commonalities in ethics, behaviors, and decision-making processes due to shared historical events. As one generation ages and a new one comes to power, decision-making patterns may be gleaned when ethics and behaviors change, and result in new decision-making processes, actions, and policies.

To illustrate, the 5th age cohort, born roughly in the 1950s, is characterized by being good stewards of the status quo. But, with such diverse backgrounds and lack of a shared future vision or values-based cohesiveness, this age cohort does not have strong bonds between them. So, whereas the previous age cohort was very driven by pride for the motherland, this group tends to fall back on old familial ties to their local hometowns.

The 6th generation, those born in the early to mid 1960s, is now in their 40s and poised to take leadership roles in about 2020. This generation is largely made up of technocrats with can-do attitudes who share a common positive vision for China's future in space. However, they will have trouble fully realizing this vision. Their shared experiences as children during the Cultural Revolution seem to have resulted in a generation that is often perceived as more politically reliable than innovative thinkers. In the 1980s, they were thus pragmatic stewards of the industry as they emerged as young adults from the Cultural Revolution. This resulted in the successes of recombinative innovation in the aerospace industry. They found they could reduce political risks, yet still make strides in the space programs by learning from foreign space programs.

Sadly, the 7th generation is often seen as China's Lost Generation of aerospace professionals. Not only do they represent the generation born into the chaos of the 1970s, they also are the first to be born under the One-Child Policy. For a variety of reasons, the central government is actively grooming younger aerospace professionals for future leaders.

This younger generation being cultivated by the current leadership in China is the 8th age cohort of aerospace professionals, born between 1980 and 1990. A lot of hope and pressure has been placed on this young class of space engineers and scientists. This is the first generation alive today to never experience civil war, foreign invasion, or mass starvation. They likely were born as middle class only-children with highly motivated intellectual parents and two sets of doting grand-parents. They consist of an imbalanced number of young males about to reach marriage age so they are professionally driven to make good salaries and attract potential wives.

The 8th generation is made up of technocrats who are innovative, creative, and uncannily confident. They likely have overseas experience, speak English, and understand China's current and future hierarchical Confucian-based place in the global space community. Their places in patron-client relationships extend beyond their chain of command by also branching out across space programs to the commercial and academic sectors, and even to international organizations. They are quite capable of building an institutional empire in aerospace technologies that could, perhaps, one day overshadow the Western aerospace industry's capacity for innovation and competition.

Chinese aerospace professionals share personal experiences through China's recent and volatile history from celebrating the founding of the PRC, surviving the Great Leap Forward, enduring the chaos of the Cultural Revolution, and experiencing the transformation of the economy, to embracing the technological revolution. Understanding the aerospace age cohorts can be used as one of several tools to better identify and understand aerospace professionals' potential decision-making trends.

9. *Domestic Influences in the Aerospace Industry*: A unique Chinese space culture has emerged. Just as ESA or the U.S. Air Force have their own distinct space cultures, so too do the components of China's domestic aerospace industry. Yet, it would be erroneous to presume Chinese space culture is remotely similar to that of Western space-faring entities. When looking at the culture of China's aerospace industry, historical patterns define different stages of cultural influences in China's recent history. China moved from a newborn state, focused on national identity and pride, to one focused on political correctness as defined by the CCP. It then shifted to a nation fixated on economic endeavors, and again focused on national identity and cultural heritage. Currently, China is focused on long-term future-oriented prosperity. By seeing China's space culture emerge during these historical stages, it is then possible to address how cultural influences may impact the aerospace industry in the future.

Just as Chinese culture has permeated the aerospace industry, so, too, has China's aerospace industry impacted Chinese culture. Beyond satellite applications to help rescue and feed people, satellite applications are also responsible for creating many life conveniences and entertainment for the Chinese people. The aerospace industry continues to make inroads with the people, for example through space science fiction, to promote further support from the Chinese people.

Because differences in language, history, and culture abound between China
and the rest of the world, it is important to be sensitive to China's space culture to
understand how aerospace activities may act as a vehicle by which Chinese people
view themselves and how space culture has affected China's concept of its place in
relation to the rest of the world. As the Chinese aerospace professionals reach
toward the Moon and eventually toward Mars, they are bringing Chinese space
culture and practices with them, which will permeate all aspects of space activities,
as it is impossible to separate culture from technology.

10. *Western Engagement with China in Space*: To state the obvious, it is critical
    that Western space-faring nations engage China on space matters at least to
    perpetuate rules of the road over a valuable and shared resource. China is
    going into space with or without foreign engagement and they will bring their
    distinct norms, culture, and codes of conduct based on the many issues pre-
    sented in this book. Therefore, more engagement between China and Western
    space leaders in sciences, business, and government should take place as has
    begun during the first Sino-U.S. Satellite Coordination Meeting, which
    recently occurred in Beijing.[8] But, what is the best way to communicate with
    leaders in China's aerospace industry? What areas of engagement would bring
    about the most fruition? First, it is important to know what China has, what it
    can do, and what it perceives its space-related needs and vulnerabilities to be.
    Second, Western space-faring communities would benefit from knowing
    China's space plans and intentions. Third, during communication and coop-
    eration, Westerners should consider specific Chinese-style negotiation tactics.
    For example, the Chinese operate within the Confucian ethic of reciprocity so
    when Western space-faring nations bestow knowledge without asking for a
    quid pro quo arrangement with China, the Chinese will tend to look at the
    information with suspicion. Working with China's method of reciprocity and
    knowing that once a deal is signed then true negotiations can begin, and then
    more productive cooperation and negotiation outcomes may result.

Productive outcomes are being accomplished in the space sciences realm. Since
the 2007 ASAT test and orbital debris creation, China has worked hard to convince
the world that it now is a responsible space-faring nation by pursuing R&D of SSA
and de-orbiting capabilities.[9] China is serious about this, and the international
space sciences community has been successful in communicating with China on
this topic. This issue affects all sectors of China's domestic space industry, for
example, right down to China's acquisition cycle where engineers are building
satellites with more fuel onboard for de-orbiting operations at the satellite's end of
life to prevent further pollution of space.

---

[8] 本刊讯，"中国与美国首次主管部门间卫星网络协调会谈在北京举行 [5].
[9] For example, in Shanghai, recent work between the Czech Republic and China is looking into
an experimental laser ranging system for space debris.

So, who is shaping whom? The Western space-faring nations can learn from past space relations. The protection of what the U.S. government considers strategic technology has served to shape the direction of China's space programs. These restrictions drove China to do things differently and explore relations with Russia and Europe. Now, China and other nations are adept at building and launching satellites without any U.S. technologies onboard. Also, exclusion of China from working on the International Space Station has had repercussions with respect to the Tiangong-1 space lab and future space station. Just as Chinese aerospace leadership has not forgotten President Carter's Moon rock gift during Zbigniew Brzezinski's 1978 visit, it is feasible that the Chinese may be able to return such a gift in the near future as a signal to the United States that China made it to the Moon during a time when the United States could not. Also, there are many lessons to be learned from ProtoStar, Ltd., and the role of space law in international disputes. Much can be learned from space history and how to apply these lessons to future dealings with China.

China is on the cusp of independently innovating space technologies. Time is limited to work within the framework that currently exists. The Chinese aerospace industry seen today will look vastly different in the near future. In that new future, current negotiation potentials may be greatly curbed. Perhaps, a radically different approach to establishing positive and fruitful aerospace relations is needed, such as the establishment of a much greater Western presence at Beihang, National University of Defense Technology, or HIT. A lot of progress has been made with military educational exchanges and visits, but more long-term stays would help establish guanxi between Western and Chinese aerospace professionals. Engagement with China will take time, patience, and iterations of attempts to communicate. This book was written to suggest and provide new approaches to understanding China in space. With the tools introduced in these chapters, Westerns can better understand, better communicate, and eventually and hopefully, better cooperate with China in space.

References

1. Fang Zhongyi, Xu Jianmin, and Guo Lujun. 1997. The development of China's meteorological satellite and satellite meteorology. In *Space science in China*, ed. Hu Wenrui, 239. Amsterdam: Gordon and Breach Science Publishers
2. D., E., Brock. 2009. Science Innovation during the Cultural Revolution: Notes from the Peking Review. *Southeast Review of Asian Studies* 31: 230.
3. Guo, Huadong, and Wu, Ji, eds. 2010. Space science and technology in China: A roadmap to 2050, eds. Beijing: Beijing Science Press.
4. Information Office of the State Council of the PRC. 2011. China's space activities in 2011, 新华, 29 December 2011; Available from http://news.xinhuanet.com/english/china/2011-12/29/c_131333479.htm.
5. 本刊讯, "中国与美国首次主管部门间卫星网络协调会谈在北京举行," 中国无线电 5 (2011).

Appendices

Appendix A

Chinese Rockets

CZ-2

Image courtesy of CGWIC

Data taken from China Great Wall Industries Corporation website at http://www.cgwic.com and CASC website at http://www.spacechina.com, unless otherwise noted.

Rocket	Date	Payload(s)	Launch site	Orbit
Recent CZ-2 launches				
CZ-2C	29 Jul 2011	SJ-11/02	Jiuquan	SSO
CZ-2C	18 Aug 2011	SJ-11/04	Jiuquan	SSO
CZ-2F	28 Sep 2011	Tiangong-01	Jiuquan	LEO
CZ-2F	01 Nov 2011	SJ-08	Jiuquan	LEO
CZ-2D	20 Nov 2011	SY-04 and Chuangxin-01/03	Jiuquan	LEO
CZ-2C	30 Nov 2011	YG-13	Taiyuan	SSO
CZ-2D	06 May 2012	Tianhui-1/02	Jiuquan	SSO
CZ-2F/G	16 Jun 2012	SZ-09	Jiuquan	LEO

Notes

- Launch capability for CZ-2C to LEO is 3,850 kg and to SSO is 1,400 kg; for CZ-2D to SSO is 1,300 kg.
- The CZ-2F is used for the manned missions and can be discerned by the telltale launch escape system on the top of the rocket.
- A CZ-2 rocket experienced a failure on 18 August 2011 when the CZ-2C experienced a malfunction between the servomechanism and the second stage Vernier engine, marking the first failure for this rocket in decades. The payload, a Shijian-11, was lost.[1]
- The CZ-2 was contracted by Motorola, Inc. in 1993 to launch multiple Iridium satellites into LEO. Between 1997 and 1999, seven CZ-2 rockets successfully launched 14 Iridium satellites into orbit.

---

[1] "Malfunction at Devices Connection Blamed for Orbiter Launch Failure," 5 Sep 2011, Beijing, from http://news.xinhuanet.com/english2010/sci/2011-09/06/c_131100892.htm.

Image courtesy of CGWIC.
Schematics for CZ-2D. **1**
Payload Fairing. **2** Payload,
3 Supporting Bay, **4** Front
Shell of Second Stage
Oxidizer Tank, **5** Equipment
Bay, **6** Inter-stage Section,
7 Second Stage Oxidizer
Tank, **8** Inter-tank Section,
9 Second Stage Fuel Tank,
10 Second Stage Vernier
Engine, **11** Second Stage
Main Engine, **12** Inter-stage
Shell Section, **13** Inter-stage
Strut Structure **14** First Stage
Oxidizer Tank, **15** Inter-tank
Section, **16** First Stage Fuel
Tank, **17** Rear Transition
Section, **18** Stabilizing Fin,
19 First Stage Engine

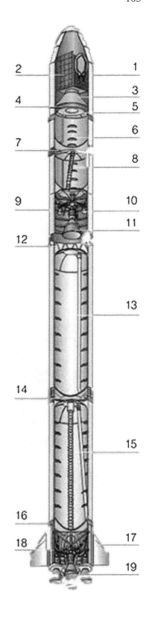

CZ-3

Image courtesy of CGWIC

Rocket	Date	Payload(s)	Launch site	Orbit
Recent CZ-3 launches				
CZ-3A	27 Jul 2011	Beidou	Xichang	IGSO
CZ-3B	12 Aug 2011	Paksat-1R	Xichang	GTO
CZ-3B	19 Sep 2011	Chinasat-1A	Xichang	GTO
CZ-3B	7 Oct 2011	Eutelsat-W3C	Xichang	GTO
CZ-3A	2 Dec 2011	Beidou	Xichang	GTO
CZ-3B	20 Dec 2011	Nigcomsat-1R	Xichang	GTO
CZ-3A	13 Jan 2012	FY-2F	Xichang	GTO
CZ-3C	25 Feb 2012	Beidou	Xichang	GTO
CZ-3B	31 Mar 2012	APSTAR-7	Xichang	GTO
CZ-3B	30 Apr 2012	2x Beidou	Xichang	GTO
CZ-3B	27 May 2012	ChinaSat-2A	Xichang	GTO
CZ-3C	25 July 2012	TianLian-03	Xichang	GTO

Notes

- Whereas the CZ-2 is used for LEO and SSO launches, the CZ-3 is used for GTO launches.
- The CZ-3B, designed for launching heavy GEO communications satellites for the international satellite launch market, is China's most powerful with four strap-on boosters, compared to the CZ-3C with two strap-on boosters.
- Launch capacity for the CZ-3A is 2,600 kg; the CZ-3B is 5,100 kg (up to 5,500 kg onboard the modified CZ-3B/E); and the CZ-3C is 3,800 kg.
- For comparison, the forthcoming CZ-5 heavy lift will be able to launch a 14 ton payload into GTO or two large GEO communications satellites thus allowing for more opportunities in the international COMSAT market.
- On 31 August 2009, a CZ-3B experienced an anomaly in the third stage during launch of Indonesia's Palapa-D satellite. Thales Alenia Space was forced to perform several satellite maneuvers to place Palapa-D in the correct orbit, which wasted spacecraft fuel and shortened the lifespan from 17 years to 10+ years.[2]

---

[2] Basedon Interview with Thales Alenia Space Chief Technical Officer from "Michael Fiat, Chief Technical Officer, Thales Alenia Space," 12 Oct 2009, http://www.spacenews.com/profiles/micael-fiat-chief-technical-officer-thales-aleniaspace.html.

Image courtesy of CGWIC.
Schematics of the CZ-3C.
1 Payload Fairing, **2** Payload,
3 Payload Adapter, **4** Vehicle
Equipment Bay, **5** LH2 Tank,
6 LOX Tank, **7** Inter-stage
Section, **8** 3rd Stage Engine,
9 Oxidizer Tank, **10** Inter-
tank Section, **11** Fuel Tank,
12 Second Stage Vernier
Engine, **13** Second Stage
Engine, **14**. Inter-stage Truss,
15 Oxidizer Tank, **16** Fuel
Tank, **17** First Stage Engine,
18 Strap-on Booster Cone,
19 Strap-on Booster Oxidizer
Tank, **20** Strap-on Booster
Fuel Tank, **21** Strap-on
Booster Engine.

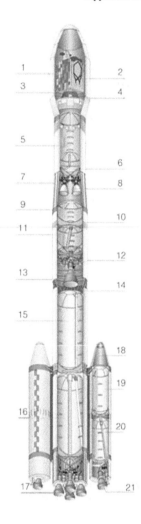

CZ-4

Image courtesy of Global Security

Rocket	Date	Payload(s)	Launch site	Orbit
Recent CZ-4 Launches				
CZ-4B	16 Aug 2011	Haiyang-02	Taiyuan	SSO
CZ-4B	9 Nov 2011	YG-12 and Tianxun-01	Taiyuan	SSO
CZ-4B	22 Dec 2011	Ziyuan-01/2C	Taiyuan	SSO
CZ-4B	9 Jan 2012	Ziyuan-03 and Vesselsat-02	Taiyuan	SSO
CZ-4B	10 May 2012	YG-14 and Tiantuo-01	Taiyuan	SSO
CZ-4C	29 May 2012	YG-15	Taiyuan	SSO

Notes

- Launch capacity to SSO for both the CZ-4B and -4C is 2,800 kg; however, the CZ-4C can house a larger payload.
- Since 1988, the CZ-4 has had 22 launches at a success rate of 100%.
- The CZ-4 launched 6x Yaogan remote sensing satellites since 2006 and 7x Shijian satellites as single and dual payloads since 2004.
- A CZ-4 launched the Fengyun-lC meteorological satellite, which was later used as a target for China's 2007 ASAT test.

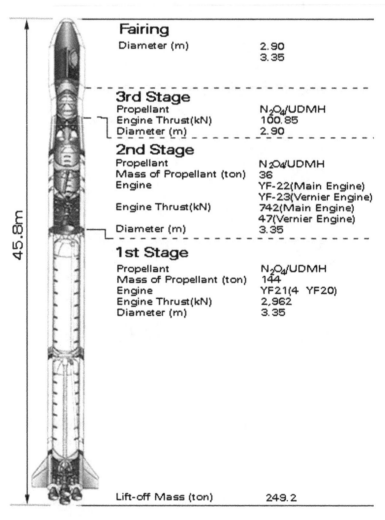

Fairing	
Diameter (m)	2.90
	3.35
3rd Stage	
Propellant	N₂O₄/UDMH
Engine Thrust(kN)	100.85
Diameter (m)	2.90
2nd Stage	
Propellant	N₂O₄/UDMH
Mass of Propellant (ton)	36
Engine	YF-22(Main Engine)
	YF-23(Vernier Engine)
Engine Thrust(kN)	742(Main Engine)
	47(Vernier Engine)
Diameter (m)	3.35
1st Stage	
Propellant	N₂O₄/UDMH
Mass of Propellant (ton)	144
Engine	YF21(4 YF20)
Engine Thrust(kN)	2,962
Diameter (m)	3.35
Lift-off Mass (ton)	249.2

Image courtesy of Global Security

Appendix B

Current On-Orbit Operational Chinese Satellites

Satellite mission	Count	Note
Communications	18	Excludes any international or foreign commercial communications companies from which China leases bandwidth (such as INMARSAT[1]).
Data relay	3	With three operational data relay satellites, China has completed its first data relay network system. The Tianlian satellites support ground fixed and mobile stations and Yuanwang SESSs for global NRT communications needs.[2]
Position, navigation, and timing	15	China plans to launch three additional Beidou satellites in 2012 to reach 30 satellites by 2020. China's goal is to establish a GNSS and break China's dependence on U.S. GPS.
Earth observation	38	This total count includes Fengyun weather satellites.
Scientific/research	17	There remains debates by Westerners as to how many of these satellites may be used for operational Earth observation missions.
Manned	2	Shenzhou-X is expected to launch by late 2012 or early 2013 and will conduct a manned rendezvous and docking with Tiangong-1 space lab.
Total	93	China plans to launch a total of 21 satellites in 2012 bringing the total to 100+ on-orbit operational satellites.

[a] Numbers are approximate.

Sources
1. "Inmarsat pic Reports 2008 Interim Results and Inmarsat Holdings Limited Reports Second Quarter 2008 Results," from http://www.inmarsat.com/About/Investors/Press_releases/00024185.aspx?language=En&textonly=False.
2. Andrew Erickson, "Satellites Support Growing PLA Maritime Monitoring and Targeting Capabilities," China Brief Vol 11 Issue 3, 10 Feb 2011, from internet http://www.jamestown.org/singleAPno_cache=1&tx_ttnews%5Btt_news%5D=37490&tx_ttnews%5BbackPid%5D=517.

Appendix C

Ground Facilities

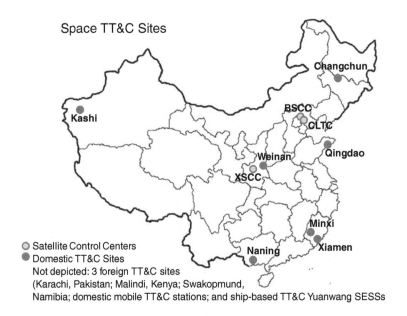

Space TT&C Sites

Changchun

BSCC
CLTC

Kashi

Weinan
Qingdao

XSCC

Minxi

○ Satellite Control Centers
● Domestic TT&C Sites
Not depicted: 3 foreign TT&C sites
(Karachi, Pakistan; Malindi, Kenya; Swakopmund,
Namibia; domestic mobile TT&C stations; and ship-based TT&C Yuanwang SESSs

Naning
Xiamen

S. Solomone, *China's Strategy in Space*, SpringerBriefs in Space Development,
DOI: 10.1007/978-1-4614-6690-1, © Stacey Solomone 2013

Space R&D and Production Centers

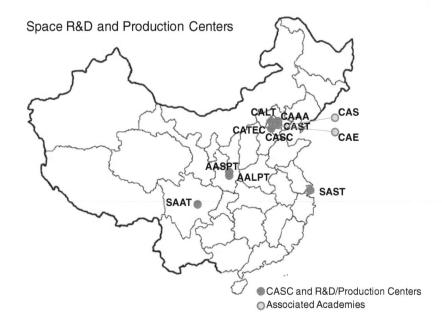

CASC and R&D/Production Centers
Associated Academies

Space Launch Centers

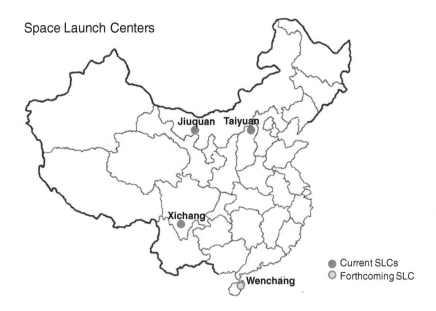

Current SLCs
Forthcoming SLC

Appendix D

Chang Zheng International Commercial Launch History

S. Solomone, *China's Strategy in Space*, SpringerBriefs in Space Development, 113
DOI: 10.1007/978-1-4614-6690-1, © Stacey Solomone 2013

No.	Payload	Launch vehicle	Customer	Launch date	Ref.
1	Microgravity test instrument	CZ-2C	MartraMaconi, France	5 Aug 1987	Piggyback
2	Microgravity test instrument	CZ-2C	Intospace, Germany	5 Aug 1988	Piggyback
3	AsiaSat-1	CZ-3	AsiaSat, HK	7 April 1990	Dedicated
4	BADR-A/ Aussat Dummy Payload	CZ-2E	SUPARCO, Pakistan	16 July 1990	Piggyback
5	Aussat-B1	CZ-2E	Aussat, Australia	14 Aug 1992	Dedicated
6	Freja	CZ-2C	SSC, Sweden	6 Oct 1992	Piggyback
7	Optus-B2	CZ-2E	Aussat, Australia	21 Dec 1992	Dedicated
8	APSTAR-I	CZ-3	APT, HK	21 July 1994	Dedicated
9	Optus-B3	CZ-2E	Optus, Australia	28 Aug 1994	Dedicated
10	APSTAR-II	CZ-2E	APT, HK	26 Jan 1995	Dedicated
11	AsiaSat-2	CZ-2E	AsiaSat,HK	28 Nov 1995	Dedicated
12	EchoStar-1	CZ-2E	EchoStar, USA	28 Dec 1995	Dedicated
13	INTELSAT-7A	CZ-3B	INTELSAT	15 Feb 1996	Dedicated
14	APSTAR-IA	CZ-3	APT, HK	3 July 1996	Dedicated
15	ChinaSat-7	CZ-3	ChinaSat, China	18 Aug 1996	Dedicated
16	Microgravity test instrument	CZ-2D	Marubeni Corp., Japan	20 Oct 1996	Piggyback
17	MabuhaySat	CZ-3B	Mabuhay, Philippines	20 Aug 1997	Dedicated
18	APSTAR-IIR	CZ-3B	APT, HK	17 Oct 1997	Dedicated
19	Iridium	CZ-2C	Motorola, USA	8 Dec 1997	Dual
20	Iridium	CZ-2C	Motorola, USA	26 Mar 1998	Dual
21	Iridium	CZ-2C	Motorola, USA	2 May 1998	Dual
22	ChinaStar-1	CZ-3B	China Orient, China	30 May 1998	Dedicated
23	SinoSat-1	CZ-3B	SinoSat, China	18 July 1998	Dedicated
24	Iridium	CZ-2C	Motorola, USA	20 Aug 1998	Dual
25	Iridium	CZ-2C	Motorola, USA	19 Dec 1999	Dual
26	Iridium	CZ-2C	Motorola, USA	12 Jun 1999	Dual
27	CBERS-01	CZ-4	INPE, Brazil	14 Oct 1999	Dedicated

(continued)

(continued)

No.	Payload	Launch vehicle	Customer	Launch date	Ref.
28	SACI	CZ-4	INPE, Brazil	14 Oct 1999	Piggyback
29	CBERS-02	CZ-4	INPE, Brazil	21 Oct 2003	Dedicated
30	APSTAR-VI	CZ-3B	APT, HK	12 Apr 2005	Dedicated
31	NigComSat-1	CZ-3B	NSRDA, Nigeria	14 May 2007	Dedicated
32	ChinaSat-6B	CZ-3B	ChinaSat, China	5 July 2007	Dedicated
33	CBERS-02B	CZ-3B	INPE, Brazil	19 Sep 2007	Dedicated
34	ChinaSat-9	CZ-3B	ChinaSat, China	9 Jun 2008	Dedicated
35	VeneSat-1	CZ-3B	MOST, Venezuela	30 Oct 2008	Dedicated
36	PALAPA-D	CZ-3B	PT IndonisiaTbk	31 Aug 2009	Dedicated
37	Paksat-1R	CZ-3B	SUPARCO	12 Aug 2011	Dedicated
38	W3C	CZ-3B	Eutelsat	07 Oct 2011	Dedicated
39	NigComSat-1R	CZ-3B	NSRDA, Nigeria	20 Dec 2011	Dedicated
40	CBERS-3	CZ-4B	INPE, Brazil	9 Jan 2012	Dual
41	Vesselsat-2	CZ-4B	LuxSpace, Luxemburg	9 Jan 2012	Dual
42	APSTAR-VII	CZ-3B	APT, HK	31 Mar 2012	Dedicated

[a] A total of 41 international satellites comprising 35 commercial launches and 6 piggyback missions. [b] Figures courtesy of CGWIC.

Glossary of Terms and Acronyms

AALPT Academy of Aerospace Liquid Propulsion Technology

AASPT Academy of Aerospace Solid Propulsion Technology

ASAT Anti-satellite

AsiaSat Asia Satellite Telecommunications Company, Ltd

ATMs Automated Teller Machines

BACC Beijing Aerospace Command and Control Center

Beihang Beijing University of Aeronautics and Astronautics

BLOS Beyond line of sight

BURO Beijing UFO Research Organization

CAAA China Academy of Aerospace Aerodynamics

CAE China Aerospace Enterprise

CALT Chinese Academy of Launch Vehicle Technology

CAMEC China Aerospace Machinery and Electronics Corporation

CAS Chinese Academy of Sciences

CASC China Aerospace Science and Technology Corporation

CAST China Academy of Space Technology

CATEC China Aerospace Times Electronics Corporation

CBERS China-Brazil Earth Resources Satellite

CCP Chinese Communist Party

CCPCC Chinese Communist Party Central Committee

CCTV China Central Television

CGWIC China Great Wall Industry Corporation

CITIC China International Trust and Investment Corporation

CLEP China Lunar Exploration Program

CLTC China Satellite Launch and Tracking Control General

CNSA China National Space Administration

COSTIND Commission on Science, Technology, and Industry for National Defense

CPPCC Chinese People's Political Consultative Conference

CZ Chang Zheng or Long March rocket

DFH Dong Fang Hong or East is Red

EOS Earth observing system

ESA European Space Agency

EVA Extra vehicular activity

FAST 500 m aperture spherical telescope

GEO Geostationary orbit

GMES Global Monitoring for Environment and Security

GNSS Global Navigation Satellite System

HADR Humanitarian Assistance and Disaster Relief

ICAO International Civil Aviation Organization

IPTV Internet Protocol Television

ISR Intelligence, Surveillance, and Reconnaissance

ITU International Telecommunication Union

MFA Ministry of Foreign Affairs

MOST Ministry of Science and Technology

NASA National Aeronautics and Space Administration

NDRC National Development and Reform Commission

NOAA National Oceanic and Atmospheric Administration

OOS On-Orbit Servicing

ORS Operationally Responsive Space

PLA People's Liberation Army

PRC People's Republic of China

R&D Research and Development

RMB Renminbi

S&T Science and Technology

SAAT Shanghai Academy of Aerospace Technology

SAST Shanghai Academy of Spaceflight Technology

SASTIND State Administration on Science, Technology, and Industry for National Defense

SBSP Space-Based Solar Power

SLC Space Launch Center

SOEs State-Owned Enterprises

SSA Space Situational Awareness

SSTL Surrey Satellite Technology, Ltd

SZ Shenzhou or Divine Capsule

TT&C Tracking, Telemetry, and Control

VSATs Very Small Aperture Terminals

XSCC Xi'an Satellite Monitor and Control Center